The Basic Book of

Drafting

Paul I. Wallach

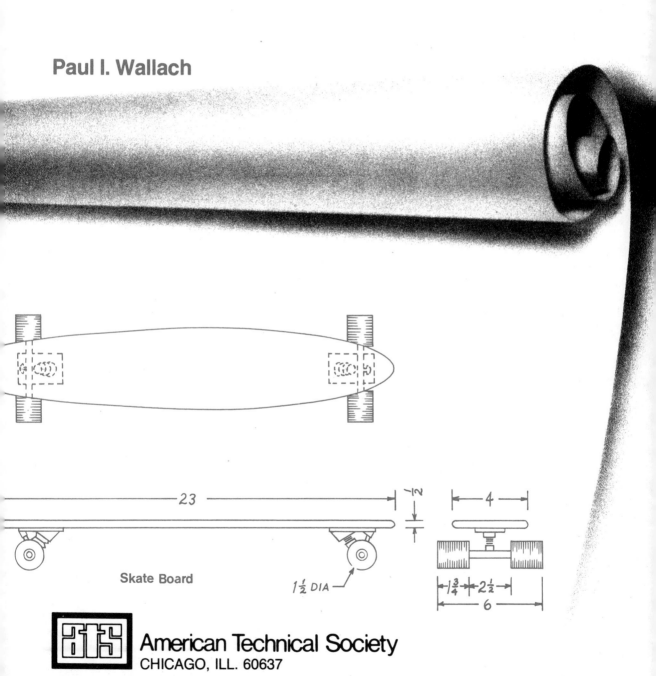

23

$\frac{1}{2}$

4

Skate Board

$1\frac{1}{2}$ DIA

$1\frac{3}{4}$ — $2\frac{1}{2}$

6

American Technical Society
CHICAGO, ILL. 60637

CONTENTS

PREFACE

THE BASIC BOOK OF DRAFTING is part of an integrated series of Industrial Arts textbooks designed to teach basic skills to beginning students. Its major objectives are career exploration, developing visual perception, knowledge of industry, and fundamental drafting abilities. The philosophy of THE BASIC BOOK OF DRAFTING is based on a recent, nationwide survey in which drafting teachers at all levels were asked to outline the courses they actually taught and let us know what types of instructional materials they actually needed. The result is a highly visual text with a controlled reading level that will help insure student success.

The author and the publisher wish to acknowledge and thank the following individuals, agencies, and corporations for their assistance and cooperation: Auto-Trol Corp., Pete Blair, Brodhead-Garrett Corp., Bruning Division of Addressograph-Multigraph, Cessna Aircraft Co., John Deere Co., Mike Giles, Modulux Division of U.S. Gypsum, NASA, Bill Rehlaender, Teledyne-Post, Vemco, and Laslo Virag, White Motors.

In line with a recent United States Bureau of Standards ruling, metre and litre are spelled meter and liter in this book.

INTRODUCTION

Imagine you have been granted a fantastic wish. You may have your own private home in a space station, built just the way you want it. Would you like a special living area, space observation decks, special rooms, your own laboratory?

You can have the space station your own way if you can show the people who build it exactly what you want.

That is what drafting does. Drafting is a way of drawing ideas so they can be made into something. An idea such

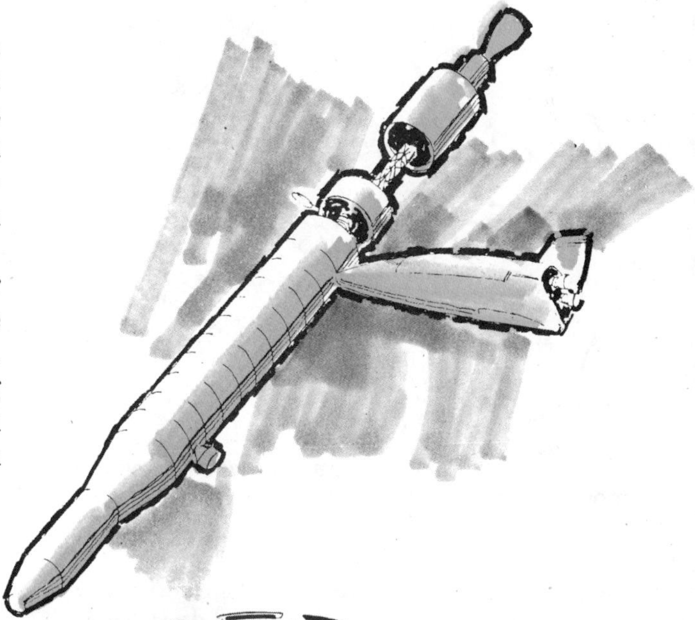

Figure I-1: Could this be your spaceship?

Figure I-2: Imagine the view of earth.

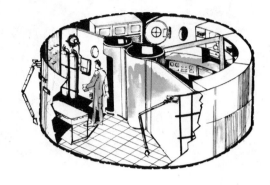

Figure I-3: Your space laboratory would need the most advanced computers available.

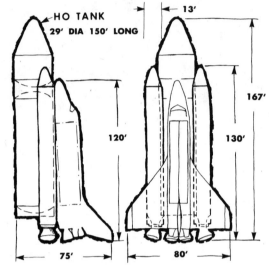

SPACE SHUTTLE SYSTEM
PARALLEL BURN

Figure I-4: A working drawing of the space shuttle.

as your space station must be drawn in many ways. Thousands of drawings would be needed to show construction details. Hundreds of drawings will be needed to show how it will look—inside and outside.

Figure I-4 is an example of a drawing that a worker will use to build the shuttle. It is called a working drawing. It shows details workers need to build the parallel burn system. Figure I-5 is an example of a perspective drawing. It could be used for reports or public relations work.

Figure I-5: A perspective drawing of the space shuttle.

All industries use drafters in their work. Space engineers, car manufacturers, house builders, and toy makers would be helpless without mechanical drawings. Without drawings, machines like airplanes (figure 1-1), trucks (figure 1-2), and space shuttles (figure 1-3) couldn't be built.

Drafting is basic to planning and making everything from a simple door stop (figure 1-4) to a Viking lander (figure 1-5). Besides showing what the

Figure 1-1: Without working drawings many machines couldn't be built.

Figure 1-3: To build something complex, like a space shuttle, takes thousands of drawings.

Figure 1-2: Trucks start as drawings on paper.

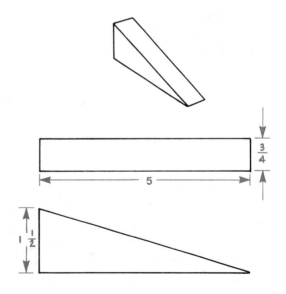

Figure 1-4: Even the simplest object, the door stop, begins as a drawing.

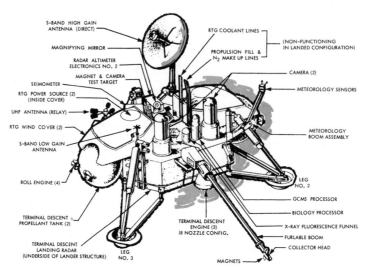

S-BAND HIGH GAIN
ANTENNA (DIRECT)

RTG COOLANT LINES

(NON-FUNCTIONING
IN LANDED CONFIGURATION)

MAGNIFYING MIRROR

PROPULSION FILL &
N₂ MAKE UP LINES

RADAR ALTIMETER
ELECTRONICS NO. 2

CAMERA (2)

MAGNET & CAMERA
TEST TARGET

METEOROLOGY SENSORS

SEIMOMETER

RTG POWER SOURCE (2)
(INSIDE COVER)

UHF ANTENNA (RELAY)

RTG WIND COVER (2)

METEOROLOGY
BOOM ASSEMBLY

S-BAND LOW GAIN
ANTENNA

ROLL ENGINE (4)

LEG
NO. 2

GCMS PROCESSOR

BIOLOGY PROCESSOR

TERMINAL DESCENT
PROPELLANT TANK (2)

X-RAY FLUORESCENCE FUNNEL

TERMINAL DESCENT
ENGINE (3)
18 NOZZLE CONFIG.

FURLABLE BOOM

TERMINAL DESCENT
LANDING RADAR
(UNDERSIDE OF LANDER STRUCTURE)

COLLECTOR HEAD

LEG
NO. 3

MAGNETS

Figure 1-5: A drawing of the Viking Lander.

Figure 1-6: The drafter is the key to a good drawing.

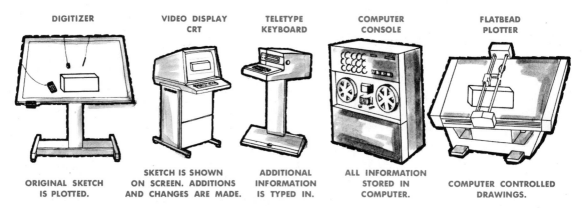

DIGITIZER	VIDEO DISPLAY CRT	TELETYPE KEYBOARD	COMPUTER CONSOLE	FLATBEAD PLOTTER
ORIGINAL SKETCH IS PLOTTED.	SKETCH IS SHOWN ON SCREEN. ADDITIONS AND CHANGES ARE MADE.	ADDITIONAL INFORMATION IS TYPED IN.	ALL INFORMATION STORED IN COMPUTER.	COMPUTER CONTROLLED DRAWINGS.

Figure 1-7: Modern technology has built computer drafting systems. A drafter had to draw the plans.

finished product will look like, mechanical drawings have notes on how to build the product. Skilled people, using the drawings and notes, can turn ideas into products.

Drafters are only one source of mechanical drawing (figure 1-6). The drawing skill of people is important, but computers also make mechanical drawings.

A computer system can store all the information about a drawing, display the drawing on a screen and then make the drawing when needed. Computer drawings can be changed immediately. Figure 1-7 is a typical computer drafting system. Such a system produces computer drawings like the one in figure 1-8.

Computers can also show different views of the same drawing (figure 1-9). A computer can design a tool, create the drawings to build it, and then run the machinery to make the tool.

Microfilm is another aid to the drafter. It is a small strip of film usually 35

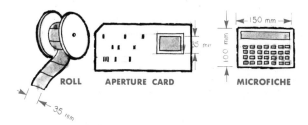

Figure 1-10: Microfilm comes in many shapes. But all of them save space.

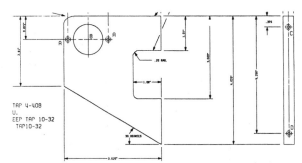

Figure 1-8: This drawing was made by a computer.

Figure 1-9: Computers can redraw an object in any position.

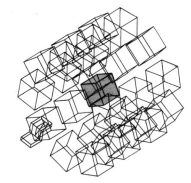

Figure 1-11: Modern technology needs so many drawings that storage is difficult.

Figure 1-12: Microfilm and microfilm readers solve some drafting storage problems.

millimeters wide (figure 1-10). Other microfilm widths are 16 mm and 105 mm. Because some projects take hundreds and even thousands of drawings, storage of the mechanical drawings is a problem (figure 1-11).

On microfilm a drawing is reduced to the size of your fingernail. When needed, a full-sized copy of the drawing can be made. The advantage of microfilm is that it can be stored in a small space and kept ready to read or to print at any time (figure 1-12).

Videotape can also be used to store drawings. The drawing can then be shown on a television screen with a voice giving descriptions and instructions. If needed, a print of the drawing can be made from the videotape.

SELF CHECK

1. List all of the industries you can think of that use drafters.
2. List three advantages of computer drawings.
3. Why is microfilm important to the drafter?
4. List three ways drawings can be stored.

Suppose you are the president of Fly 'Em High Space Ship Corporation. Your company is making the first space ship to go to Pluto. Would you say to your thousands of employees, "OK, you all know what a spaceship is. Get to work."? Of course you wouldn't. Different people have special jobs only they can do. Each must be given their own part of the space ship project to do (figure 2-1).

Drafters may also have special jobs.

The <u>tracer</u> is a person who traces or copies drawings another drafter has made. Many highly skilled drafters began as tracers. Being a tracer is an entry level job. That means the tracer doesn't need as many skills as some of the other drafters. <u>Detailers</u> make simple changes on existing drawings. A senior detailer, who has more knowledge and experience than a junior detailer, may make a complete drawing from a design or from verbal directions.

<u>Checkers</u> look for errors in other people's drawings. The professional checker must have lots of experience and know a great deal about drafting.

A <u>designer</u> is a problem solver who uses drafting skills, creativity, and technical information to help complete a design. The designer may work with engineers and scientists to create the right design.

Besides the tracer, detailer, checker, and designer, there are other specialized drafters. Each specialty depends on different skills and responsibilities, but each makes the finished project possible.

Specialty drafting skills are used in many different industries. The drafter must learn the special skills for the industry they want to work in.

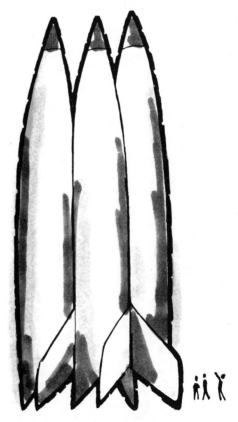

Figure 2-1: Spaceships challenge the drafter's ability.

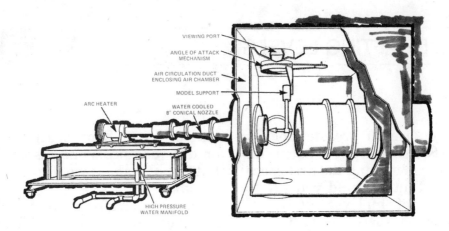

Figure 2-2: Technical illustrations may be a simple cartoon or an atomic energy complex.

Commercial artists usually work in advertising. They must know about art materials and printing. Commercial artists use oils or acrylics, charcoal, water colors, and India ink.

Technical illustrators usually draw in ink. Ink drawings reproduce well under all conditions. A technical illustrator may draw everything from technical drawings to cartoons.

Tool designers need to know machines and engineering, manufacturing methods, and mathematics as well as sketching and drawing.

An architect designs buildings by considering how people will use them and how they will look. Like all drafters, they must know the basic drawing skills. But they must also know building methods, materials, and artistic design.

SELF CHECK

1. What does a tracer do?
2. What is a person who looks for drafting errors called?
3. What does a designer do?
4. List three "specialty drafting" areas.

STARTING TO DRAW

2

Before you start to draw, there are several areas that you must know about and certain skills that you should have. These areas are:
- Lettering
- Drawing equipment and supplies
- Line work
- Sheet layout

When you understand and are able to perform these skills, you will be ready to start with your mechanical drafting.

ABCDEF GHIJKL M N
OP QRS TU VWX YZ
1 2 3 4 5 6 7 8 9 0 &

Figure 3-1: Vertical single stroke gothic letters and numbers.

ABCDEF GHI JKLMNO

PQRSTUVWXYZ

1234567890 .

Figure 3-2: Microfont letters and numbers.

ABCDEFGHIJKLMN
OPQRSTUVWXYZ
1234567890&

Figure 3-3: Slanted single stroke gothic letters and numbers.

Good lettering is as important as good drafting. Clear letters on a drawing help others to read your drawings. Most employers will not hire you if your lettering is poor.

Two types of lettering are used on most mechanical drawings. These two methods of lettering are:
● Single stroke gothic (figure 3-1)
● Microfont (figure 3-2)

Single stroke gothic letters have been used for many years. Microfont lettering was invented for drawings that are recorded on microfilm, a fairly new process. When microfilmed and then reprinted, single stroke gothic letters and numbers lack clarity.

Single stroke gothic letters can be printed straight up and down (vertically) as in figure 3-1. Single stroke gothic can also be slanted as in figure 3-3. Microfont lettering is always straight up and down. Any method that correctly forms single stroke gothic or microfont letters and numbers will be satisfactory.

Figure 3-4: Lettering is done between two sharp, thin, light lines called guide lines.

Guide lines (figure 3-4) help keep letters and numbers the same height. Before you start to letter, you must draw guide lines with a sharp 4H or 6H pencil lead.

Guide lines are set 1/8″ apart because most lettering is 1/8″ high, but fractions need more room. Fractions are usually 3/16″ high (1/8″ equals 2/16″). Each number should be slightly smaller than 1/8″. The fraction bar is straight and the numbers do not touch the bar (figure 3-5).

Bigger letters are sometimes used for titles or on very large drawings. Guidelines for such lettering are usually drawn 3/16″ or 1/4″ apart (figure 3-6).

Figure 3-8: A piece of paper under your hand will help prevent smudging.

Figure 3-5: Fraction heights for 1/8″ numbers are 3/16″ to 1/4″ high.

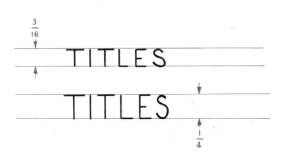

Figure 3-6: Titles are larger than regular lettering. Large drawings may also use larger lettering.

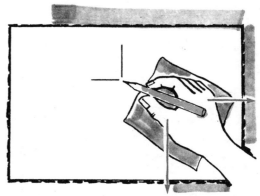

Figure 3-9: Pull the pencil. Do not push it to draw.

Figure 3-7: Correct and incorrect lettering.

BASIC DRAFTING	CORRECT
BASIC DRAFTING	TOO CLOSE
BASIC DRAFTING	TOO FAR
BASIC DRAFTING	HEIGHTS DIFFERENT

Only practice will give you the "feel" for making letters look right. Figure 3-7 shows correct and poor examples of lettering styles. Figures 3-8 through 3-13 show examples of good practice rules and some practice strokes.

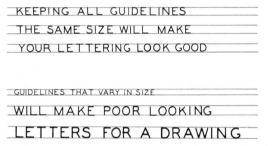

Figure 3-10: Keep all guide lines the same width. Measure the guidelines in both samples.

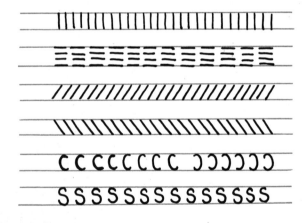

Figure 3-11: Practice exercise for vertical lettering.

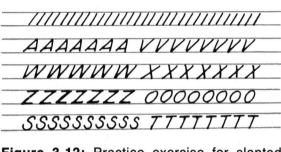

Figure 3-12: Practice exercise for slanted lettering.

Figure 3-13: Note how the vertical (colored) strokes are drawn first.

SELF CHECK

1. What are the two lettering styles used on most mechanical drawings?
2. Which lettering style is used on drawings that will be microfilmed?
3. How high should most letters be?
4. How high should fractions be?

Problems for Unit 3

VERTICAL SINGLE STROKE GOTHIC

ABCDEFGHIJKLMNOPQRSTUVWXYZ   1234567890 $\frac{1}{2}$ $\frac{3}{4}$ $\frac{7}{8}$

SLANTED SINGLE STROKE GOTHIC

ABCDEFGHIJKLMNOPQRSTUVWXYZ   1234567890 $\frac{1}{2}$ $\frac{3}{4}$ $\frac{7}{8}$

MICROFONT

ABCDEFGHIJKLMNOPQRSTUVWXYZ   1234567890

Problem 3-1: This is a practice lettering sheet which you can make for yourself.

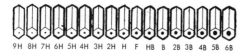

9H 8H 7H 6H 5H 4H 3H 2H H F HB B 2B 3B 4B 5B 6B

GRADES OF PENCILS USED IN DRAFTING

Figure 4-1: Grades of pencil hardness used in drafting.

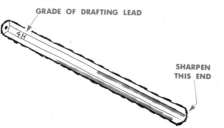

GRADE OF DRAFTING LEAD

4H

SHARPEN THIS END

Figure 4-2: Be sure to sharpen the correct end of the pencil.

Figure 4-3: A mechanical pencil sharpener.

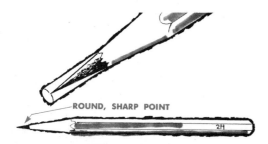

ROUND, SHARP POINT

2H

Figure 4-4: A regular sharpener cuts the wood and points the lead.

Pencils and pens are called marking media when they are used in drafting. Different grades of pens and pencils make light and dark lines. There are three kinds of pencils and two kinds of pens used in drafting:

PENCILS
- Wood encased pencils
- Lead holder pencils
- 0.5 mm automatic pencils

PENS
- Technical pens
- Ruling pens

Wood encased pencils come in 17 grades of hardness (figure 4-1). "H" numbers are hard. "B" numbers are soft. The thinner the lead, the harder it is. The thicker the lead, the softer it is (figure 4-1). The drafter uses six of these grades. Hard leads are used for light outline work. Hard leads used by the drafter are 4H, 3H and 2H. Soft

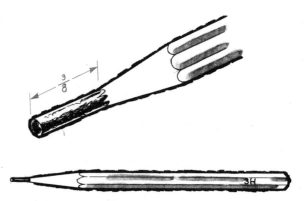

$\frac{3}{8}$

3H

Figure 4-5: A mechanical drafting sharpener removes only the wood.

14

leads darken the drawing. H, F or HB leads are used by the drafter for dark line work. Practice making marks with different pencil leads.

All drafting pencils are marked on one end (figure 4-2). Do not sharpen the end with the number. If you do, you will sharpen the marking number off.

The easiest way to sharpen your pencil is to use a pencil sharpener (figure 4-3). A regular pencil sharpener gives a sharp, round point ready for drawing (figure 4-4). A special drafting pencil sharpener takes off only the wood (figure 4-5). The lead must then be pointed with a file, sandpaper or a lead pointer (figure 4-6).

<u>Lead holders</u> (mechanical drawing pencils, figure 4-7) are pointed the same way or with a special pencil

Figure 4-8: Pointing a drafting lead.

Figure 4-9: Rotating your pencil while drawing will keep the point sharp.

Figure 4-6: Pointing the lead with a file or sandpaper.

Figure 4-7: Drafting lead holders and lead refills.

pointer (figure 4-8). After sharpening the lead, wipe the loose graphite off with a cloth.

Rotate the pencil while drawing a line (figure 4-9). If you do not rotate the pencil, the point will become a chisel point and make uneven line widths (figure 4-10).

The 0.5 mm (millimeter) <u>automatic pencil</u> has a lead so thin it does not need sharpening (figure 4-11). Leads for the 0.5 mm automatic pencil come

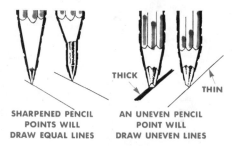

Figure 4-10: It is important to keep points round and sharp.

Figure 4-11: The 0.5 mm automatic pencil does not need sharpening.

Figure 4-13: Sets of technical pens come with points that draw different line widths.

PLACE DROP OF INK

Figure 4-14: Ruling pens must be adjusted to draw different line widths.

in all grades of hardness. But the pencil takes practice to use. The leads are so thin they snap off easily while drawing.

Technical pens (figure 4-12) are used to do drawings in black ink. Inked drawings make better copies than pencil drawings. Technical pens come in different line widths (figure 4-13). Ruling pens can be set for any line width (figure 4-14). Ruling pens don't cost much, but they take practice to use.

SELF CHECK

1. List three kinds of pencils used in drafting.
2. List three pencil hardnesses that can be used to draw dark lines
3. List three pencil hardnesses that can be used to draw light lines.
4. How can you keep a pencil sharp while drawing lines?

The drafter uses five types of drawing paper. The professional drafter must know the best kinds of drawing paper for the pencils or pens being used. There are also different ways to make copies of drawings. The drafter must select the best paper for the copy method that will be used. Paper also comes in different sizes of rolls and cut sheets (figure 5-1). Cut sheets are ordered by a letter code (figure 5-2).

The five types of drawing paper are called:
● Drawing paper
● Tracing paper
● Vellum
● Polyester film
● Graph paper

Drawing paper is a heavy paper that you cannot see through. Drawings done on drawing paper can be copied with a photographic process. Usual drawing paper colors are white, buff or green.

Tracing paper is a thin and inexpensive paper. Light can pass through tracing paper for the making of blueprints or white prints. Vellum is heavier than tracing paper, but light can still pass through it. More costly than tracing paper, vellum is stronger and will not tear as easily.

Polyester film is impossible to tear.

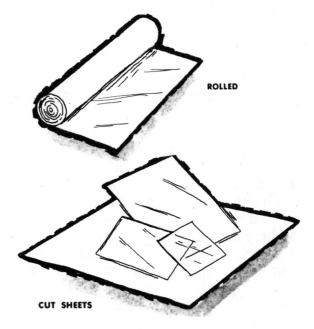

Figure 5-1: Drawing papers can be purchased in rolls or cut sheets.

A	9 x 12 OR	8 1/2 x 11
B	12 x 18 OR	11 x 17
C	18 x 24 OR	17 x 22
D	24 x 36 OR	22 x 34
E	36 x 48 OR	34 x 44
F	28 x 40 OR	28 x 40

Figure 5-2: Cut sheets of drawing paper are ordered by letter code.

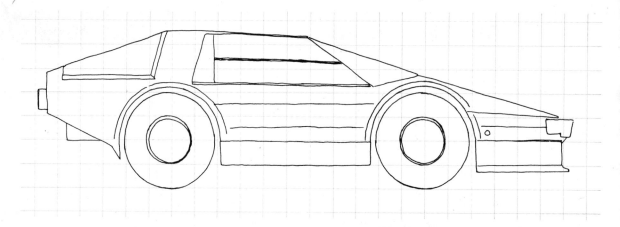

Figure 5-3: Squared 1/8″ or 1/4″ graph paper helps proportion your sketches.

Light passes through polyester film, easily and it makes very good blueprints.

Graph paper is good for freehand drawings. The paper is divided into light graph squares (figure 5-3). Use 1/8″ or 1/4″ graph squares for sketching. Graph paper comes in paper, vellum, or polyester film. The light lines on it will not print on a blueprint.

SELF CHECK

1. List two ways drafting paper is sold.
2. What kind of code is used to describe the different sizes of drafting paper?
3. How big is an "A" sheet?
4. List five different types of drafting media.

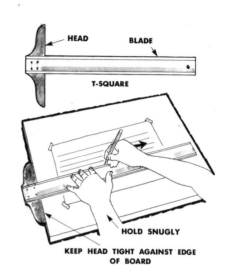

Take a look at the drafting tools shown in this unit. These are the tools you must learn to use in this drafting class.

The illustrations in this unit will show you how to use most drafting tools. Use your best tool, your mind, to examine each tool illustration, and then practice using each tool. Imagine what the tool does.

Figure 6-2: The T-square is used to draw horizontal lines.

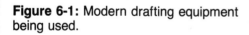

Figure 6-1: Modern drafting equipment being used.

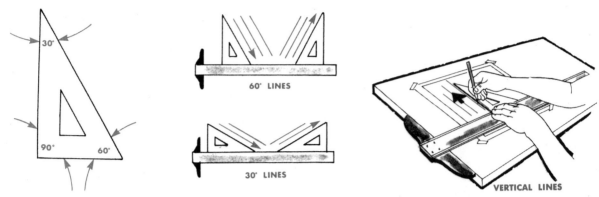

Figure 6-3: Drawing vertical, 30°, and 60° lines with a T-square and a 30°-60° triangle.

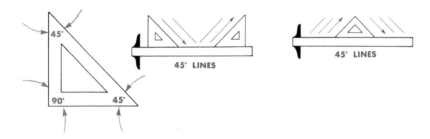

Figure 6-4: Drawing 45° lines with a T-square and a 45° triangle.

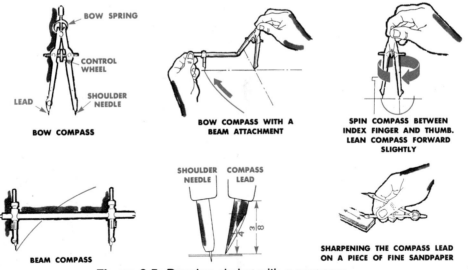

Figure 6-5: Drawing circles with a compass.

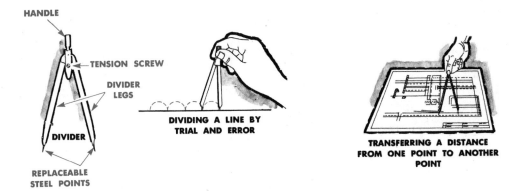

Figure 6-6: Dividers are used to divide lines or transfer distances from one point to another.

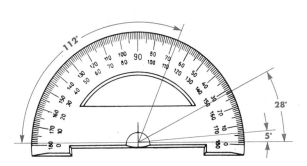

Figure 6-7: Measuring the number of degrees in an angle with a protractor.

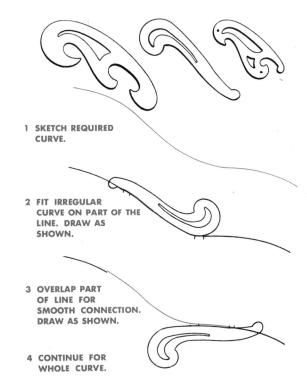

1 SKETCH REQUIRED CURVE.

2 FIT IRREGULAR CURVE ON PART OF THE LINE. DRAW AS SHOWN.

3 OVERLAP PART OF LINE FOR SMOOTH CONNECTION. DRAW AS SHOWN.

4 CONTINUE FOR WHOLE CURVE.

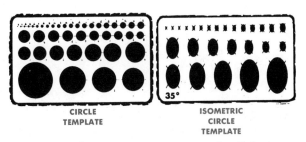

CIRCLE TEMPLATE

ISOMETRIC CIRCLE TEMPLATE

Figure 6-8: The circle and isometric circle templates can be used in beginning drafting.

Figure 6-9: Drawing a smooth curved line with an irregular curve.

Figure 6-10: A dusting brush won't smudge your drawing when removing eraser particles.

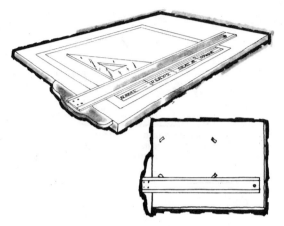

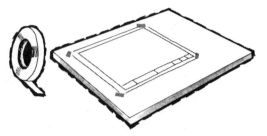

TAPE DRAWING PAPER TO UPPER LEFT SECTION OF DRAWING
BOARD. PAPER MUST BE STRAIGHT WITH T SQUARE

Figure 6-11: Drafting boards come in different sizes. The 20x26 inch board is a good size for beginning students.

Figure 6-13: Use drafting tape to hold the drawing to the board. It will not damage the board or the drawing.

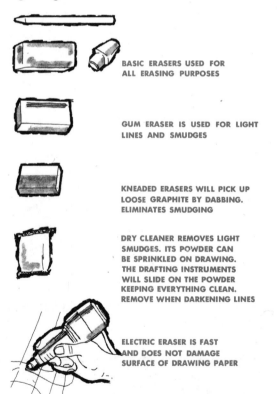

BASIC ERASERS USED FOR
ALL ERASING PURPOSES

GUM ERASER IS USED FOR LIGHT
LINES AND SMUDGES

KNEADED ERASERS WILL PICK UP
LOOSE GRAPHITE BY DABBING.
ELIMINATES SMUDGING

DRY CLEANER REMOVES LIGHT
SMUDGES. ITS POWDER CAN
BE SPRINKLED ON DRAWING.
THE DRAFTING INSTRUMENTS
WILL SLIDE ON THE POWDER
KEEPING EVERYTHING CLEAN.
REMOVE WHEN DARKENING LINES

ELECTRIC ERASER IS FAST
AND DOES NOT DAMAGE
SURFACE OF DRAWING PAPER

Figure 6-12: Different types of drawing cleaners.

Figure 6-14: An eraser shield protects lines that are not to be erased.

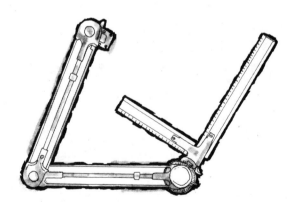

Figure 6-15: The drafting machine takes the place of the T-square, triangles, scale and protractor.

Figure 6-16: The parallel slide takes the place of the T-square.

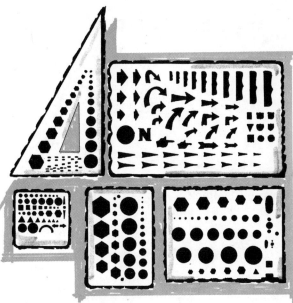

Figure 6-19: Different types of templates make the drafter's job easier.

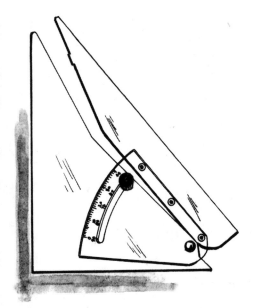

Figure 6-17: The adjustable triangle takes the place of a 30°—60° triangle, a 45° triangle, and a protractor.

LETTER GUIDE

Figure 6-18: A flexible curve can be adjusted to almost any irregular line.

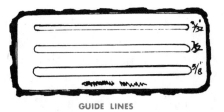

GUIDE LINES

Figure 6-20: Templates for lettering help keep your letters even.

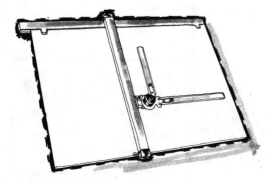

Figure 6-21: Track drafting machines are used by professional drafters.

Figure 6-24: Lettering sets make clean sharp letters.

SMALL TABLE

LARGE TABLE

Figure 6-22: Drafting tables come in many different sizes and shapes.

Figure 6-25: The drop bow compass is used to draw small circles and arcs.

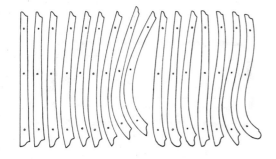

Figure 6-23: Lofting curves are used like irregular curves.

Figure 6-26: Proportional dividers make it easy to scale a drawing up or down.

Figure 6-27: A perspective drawing board makes perspective drawing much quicker and easier.

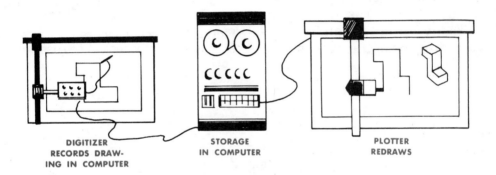

DIGITIZER
RECORDS DRAW-
ING IN COMPUTER

STORAGE
IN COMPUTER

PLOTTER
REDRAWS

Figure 6-28: Computer drafting is the wave of the future.

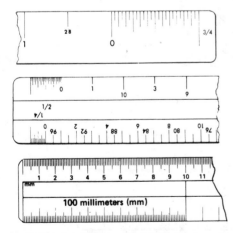

Figure 6-29: Architects scale (top), mechanical scale (center), and metric scale (bottom).

	SELF CHECK

1. Which tools are used to draw horizontal lines?
2. Which tools are used to draw vertical lines?
3. Which tools can be used to divide lines?
4. Which tools can be used to draw circles?

The mechanical scale used in drafting was developed from the measuring system we use every day. When people had a simple life, it was not important to measure carefully (figure 7-1). In early history, kings and other leaders used parts of their bodies for measurements (figures 7-2). The measuring system we use today was developed from this. Today all rulers and measuring devices must be of the exact same

Figure 7-1: Accurate measurement was not necessary when mechanical devices were simpler.

Figure 7-2: Early measuring systems were based on unequal distances.

Figure 7-3: A mechanical drafting scale is divided into inches and fractions of an inch.

length as the standard measuring rod of the Bureau of Standards and Measures.

Learning how to use a ruler and a mechanical scale properly is critical for the drafter. A mismeasurement in a drawing could cause a spaceship or aircraft to crash. At the very least, a piece that is built to the wrong size must be thrown away and done again.

Three types of measurements are used in mechanical drawing. A single drawing, however, is always done in only one drafting scale. The scales are:

● The mechanical drafting scale that uses fractions (figure 7-3).
● The engineer's drafting scale that uses decimals.
● The metric drafting scale that uses millimeters.

The typical mechanical drafting scale is one foot (12″) long. On it the inch is divided into halves, quarters, eighths, and sixteenths (figure 7-4). Sometimes scales are divided into thirty-seconds of an inch (figure 7-5). The accuracy of your drawings should be within one thirty-second of an inch. Your drafting teacher will check the accuracy of your drawing.

Often a drafter must draw an object larger or smaller than its actual size. This is called scaling a drawing up or down and will not be covered in this book. All the drawings in this book will be drawn full size. The full size scale is shown as:
SCALE: FULL SIZE or SCALE: 1:1

SELF CHECK

1. How were the first measurements made?
2. Why must measurement units be exact?
3. What units are used on a mechanical drafting scale?
4. What is a full scale drawing?

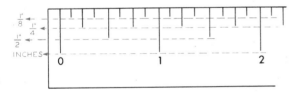

Figure 7-4: Divisions on a typical mechanical drafting scale.

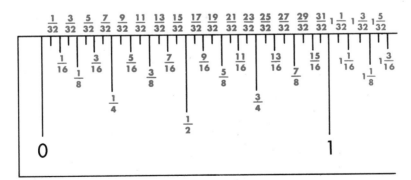

Figure 7-5: Inches are divided into thirty-seconds of an inch (1/32).

Problems for Unit 7

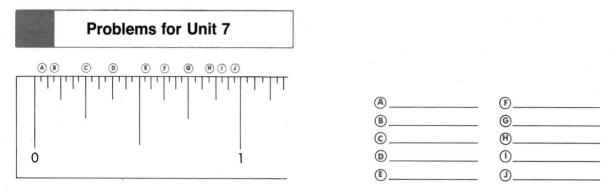

(A) _____   (F) _____
(B) _____   (G) _____
(C) _____   (H) _____
(D) _____   (I) _____
(E) _____   (J) _____

Problem 7-1: Name the fraction on the scale under each number.

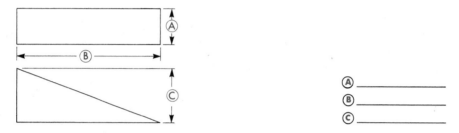

(A) _____
(B) _____
(C) _____

Problem 7-2: Use a mechanical scale to measure the distances indicated by letters on this drawing. Letter your answers.

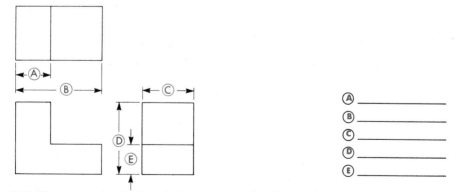

(A) _____
(B) _____
(C) _____
(D) _____
(E) _____

Problem 7-3: Use a mechanical scale to measure the distances indicated by letters on this drawing. Letter your answers.

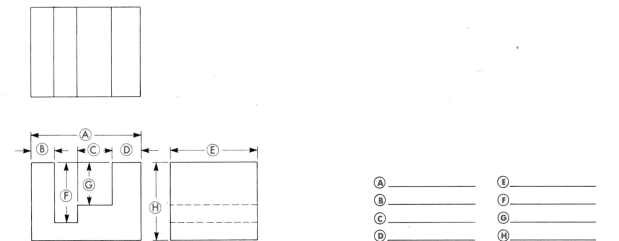

Ⓐ _____	Ⓔ _____
Ⓑ _____	Ⓕ _____
Ⓒ _____	Ⓖ _____
Ⓓ _____	Ⓗ _____

Problem 7-4: Use a mechanical scale to measure the distances indicated by letters on this drawing. Letter your answers.

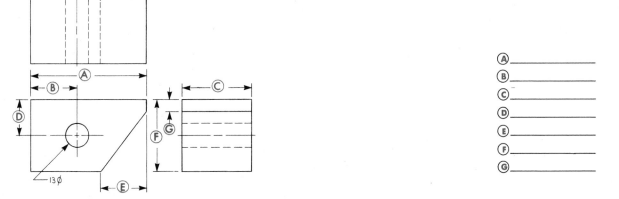

Ⓐ _____
Ⓑ _____
Ⓒ _____
Ⓓ _____
Ⓔ _____
Ⓕ _____
Ⓖ _____

Problem 7-5: Use a mechanical scale to measure the distances indicated by letters on this drawing. Letter your answers.

You must also understand the engineer's scale. More than half the drafting industry prefers the engineer's scale to the mechanical drafting scale.

The engineer's scale is a decimal scale (figure 8-1). There are no fractions. The inches are the same. But each inch is divided into ten parts (figure 8-2). Each part is one-tenth of an inch. In decimal form, one-tenth of an inch is written 0.1 inch.

The next decimal division is one-hundredth of an inch. It is written 0.01 inch (figure 8-3). One-hundredths of an inch are too small to print on a scale. To draw a hundredth of an inch, you must estimate the length.

Each tenth of an inch, shown on your scale, has ten one-hundredths of an

Figure 8-1: The engineer's scale is a decimal scale. Fractions are not used.

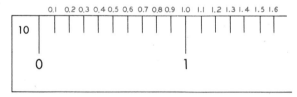

Figure 8-2: The inch is divided into ten parts on the engineer's scale.

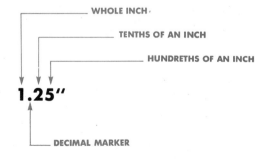

Figure 8-3: The second number behind the decimal is hundredths of an inch (1/100).

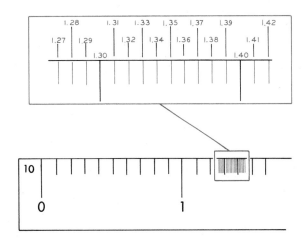

Figure 8-4: On the engineer's scale, hundredths of an inch must be estimated.

inch (figure 8-4). Divide the tenth of an inch in half and you have five one-hundredths of an inch. Slightly more than half that distance is three one-hundredths. Slightly less is two one-hundredths.

Read the dimensions in figure 8-5 to the closest one-hundredth of an inch.

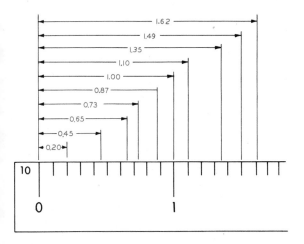

Figure 8-5: Reading the decimal scale to one-hundredth of an inch.

SELF CHECK

1. What is the smallest division printed on the engineer's scale?
2. How many units is an inch divided into on the engineer's scale?
3. How is three-tenths of an inch written in decimals?
4. How is five one-hundredths of an inch written in decimals?

Problems for Unit 8

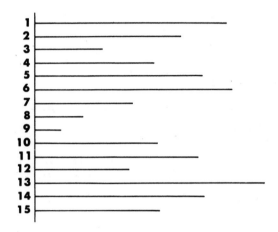

Problem 8-1: Measure these lines to the closest hundredth of an inch.

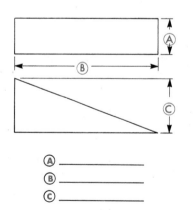

Ⓐ _____

Ⓑ _____

Ⓒ _____

Problem 8-2: Use an engineer's scale to measure the distances indicated by letters on this drawing. Letter your answers.

Ⓐ _____

Ⓑ _____

Ⓒ _____

Ⓓ _____

Ⓔ _____

Problem 8-3: Use an engineer's scale to measure the distances indicated by letters on this drawing. Letter your answers.

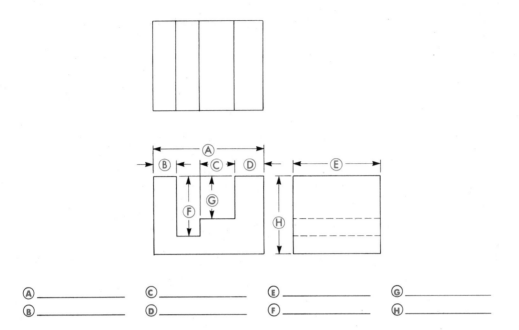

A _____   C _____   E _____   G _____
B _____   D _____   F _____   H _____

Problem 8-4: Use an engineer's scale to measure the distances indicated by letters on this drawing. Letter your answers.

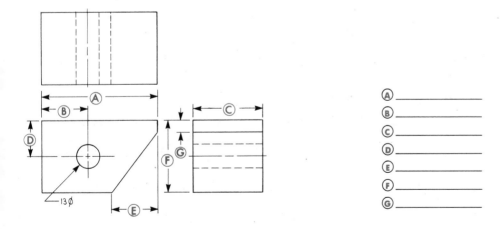

A _____
B _____
C _____
D _____
E _____
F _____
G _____

Problem 8-5: Use an engineer's scale to measure the distances indicated by letters on this drawing. Letter your answers.

METRIC MEASUREMENTS

The metric system of measurement is important because most countries in the world already use it. The United States is also converting to metrics. Start studying the metric system by learning the metric measuring terms in figure 9-1.

Measuring length in mechanical drawing starts with the <u>millimeter.</u> Milli means thousandth. There are 1 000 millimeters in a meter (figure 9-2). The symbol for millimeter is mm. Because all dimensions on a mechanical drawing are in millimeters, it is not necessary to put mm after the numbers (figure 9-3).

Measuring is simple with a metric ruler. Each line is one millimeter (figure 9-4). Count the number of millimeters you need, and you are

WHEN WE MEASURE	THE UNIT NAME IS	THE SYMBOL IS	IT TELLS
length	metre	m	how long
weight	gram	g	how heavy
temperature	degree Celsius	°C	how hot
capacity	litre	ℓ	how full

Figure 9-1: Once you know what the metric terms mean, they are easy to use.

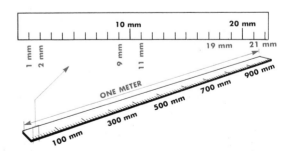

Figure 9-2: There are 1 000 millimeters in one meter.

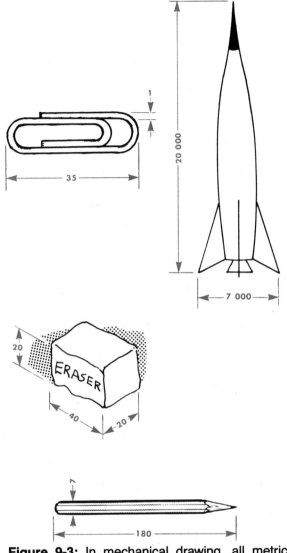

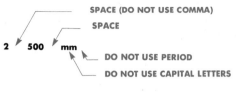

3 000 000 mm	7 mm
10 000 mm	2 mm
5 567 mm	1 mm
1 000 mm	1.5 mm
750 mm	1.25 mm
80 mm	0.75 mm
32 mm	0.03 mm

Figure 9-5: Rules and examples for writing distances in millimeters.

Figure 9-3: In mechanical drawing, all metric measurements are in millimeters.

Figure 9-6: The English measuring system uses many units of distance for measuring.

Figure 9-4: Each line is one millimeter. Each 1 000 millimeters equals one meter.

measuring in the metric system. The rules for metrics in drafting are shown in figure 9-5. Using metric measurements you do not use fractions, hundredths, tenths or the many types of measuring units we now use (figure 9-6).

You measure weight in metrics by <u>grams</u> and <u>kilograms.</u> It takes 28 grams to equal one ounce. One thousand grams equal one kilogram and one kilogram weighs 2.2 pounds (figure 9-7).

Temperature measurements in metrics are on the <u>Celsius scale.</u> To learn the Celsius thermometer, study figure 9-8.

<u>Liters</u> measure capacity in metrics. A liter is a little more than one quart.

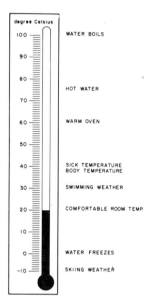

Figure 9-8: In the metric system temperature is measured in degrees celsius.

Figure 9-7: The metric unit of weight is the gram.

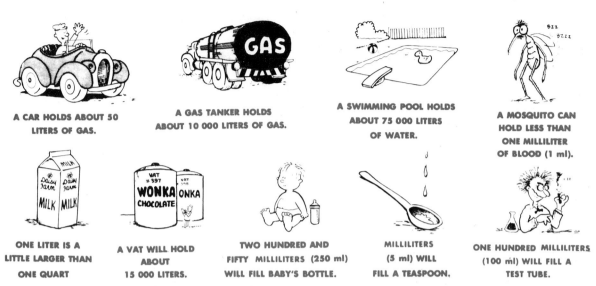

Figure 9-9: In the metric system capacity is measured in milliliters.

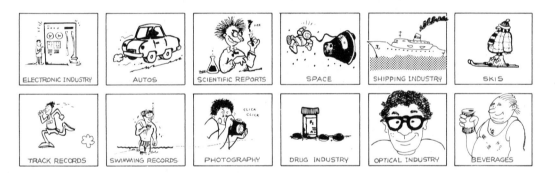

Figure 9-10: Metrics are already being used by some of the industries in the United States.

There are a thousand milliliters in a liter (figure 9-9).

Many industries in the United States now use the metric system (figure 9-10). These industries agree that the metric system is easy to learn and use.

SELF CHECK

1. What is the basic metric unit of length?
2. What is the metric unit of length used on all mechanical drawings?
3. What is the basic metric unit for measuring capacity?
4. How many millimeters are there in one meter?

Problems for Unit 9

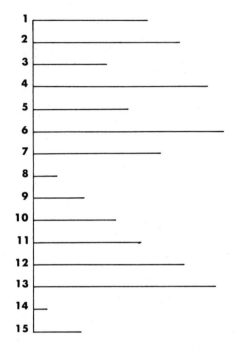

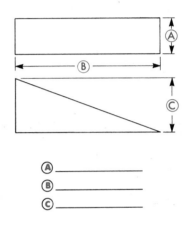

Ⓐ	_____
Ⓑ	_____
Ⓒ	_____

Problem 9-1: How many millimeters (mm) long is each line?

Problem 9-2: Use a metric scale to measure the distances indicated by letters on this drawing. Letter your answers.

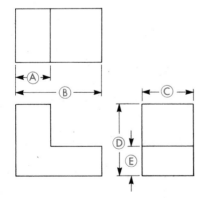

Ⓐ	_____
Ⓑ	_____
Ⓒ	_____
Ⓓ	_____
Ⓔ	_____

Problem 9-3: Use a metric scale to measure the distances indicated by letters on this drawing. Letter your answers.

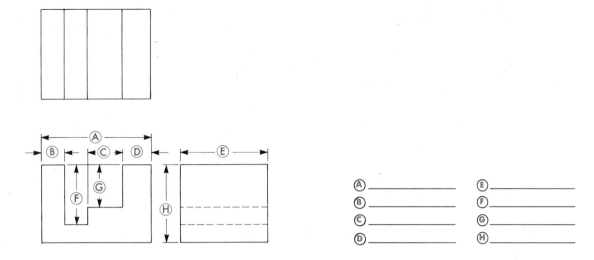

Ⓐ _____	Ⓔ _____
Ⓑ _____	Ⓕ _____
Ⓒ _____	Ⓖ _____
Ⓓ _____	Ⓗ _____

Problem 9-4: Use a metric scale to measure the distances indicated by letters on this drawing. Letter your answers.

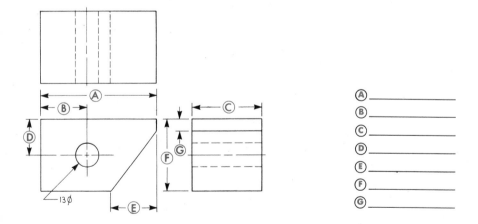

Ⓐ _____
Ⓑ _____
Ⓒ _____
Ⓓ _____
Ⓔ _____
Ⓕ _____
Ⓖ _____

Problem 9-5: Use a metric scale to measure the distances indicated by letters on this drawing. Letter your answers.

In this unit you will study the alphabet of lines, and learn how to read and draw the different types of lines used to make mechanical drawings.

Each type of line in the alphabet of lines has its own meaning and is drawn at a special width. There are three groups of lines (figure 10-1):
- Very wide lines
- Wide lines
- Thin lines

On a drawing, the thickness of each type of line must be constant (figure 10-2). At first it is hard to estimate the widths of lines as they are drawn. With practice, you will become very accurate at drawing even lines. The names and descriptions of these lines are shown in figure 10-5 along with their correct sizes and uses.

Guide lines and outline lines should not show on a finished mechanical drawing (figure 10-3). All other lines must be sharp, clear, and of the correct width (figure 10-4). All the lines in a finished mechanical drawing should be black. Do not draw fuzzy, gray lines. Black lines make better copies.

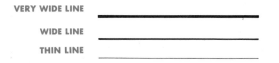

Figure 10-1: All lines on a mechanical drawing have widths that are classified as very wide, wide or thin.

Figure 10-3: Construction and guide lines do not show on a finished drawing.

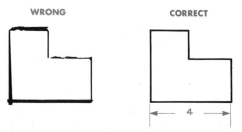

Figure 10-2: Good drawings have consistent line thicknesses.

Figure 10-4: Good lines are always sharp and black.

NAME	LINE	DESCRIPTION	SIZES	EXAMPLES
1. BORDER LINE		A VERY HEAVY LINE THAT FRAMES THE DRAWING.	A VERY WIDE SOLID LINE	
2. CUTTING PLANE LINE		A VERY HEAVY LINE THAT SHOWS WHERE A LINE CUTS THROUGH AN OBJECT TO SHOW HOW THE INSIDE LOOKS.	A VERY HEAVY BROKEN LINE. LONG DASH IS 1" SHORT DASH IS 1/4", SPACES 1/8"	
3. VISIBLE LINE		A HEAVY LINE THAT SHOWS THE FORM OF AN OBJECT. IT IS ALSO CALLED AN "OBJECT LINE".	A WIDE SOLID LINE	
4. HIDDEN LINE		A THIN LINE THAT HAS BROKEN DASHES. IT SHOWS A HIDDEN EDGE. ALSO CALLED AN "INVISIBLE LINE".	DASHES ARE 1/8" LONG WITH 1/32" SPACES.	
5. CENTER LINE		A THIN LINE WITH LONG AND SHORT DASHES. IT SHOWS THE CENTER OF OBJECTS.	DASHES ARE 1" LONG AND THE SHORT DASH 1/8" WITH 1/16" SPACES.	
6. SECTION LINE		A THIN LINE THAT SHOWS A SOLID SURFACE WHEN A CUTTING PLANE LINE PASSES THROUGH IT. ALSO CALLED CROSS-HATCHING LINES.	THIN LINE NORMALLY DRAWN AT A 45° ANGLE.	
7. LEADER		A THIN LINE WITH AN ARROWHEAD ON ONE END. ON THE OTHER END IS A DIMENSION OR NOTE.	A THIN LINE USED TO SHOW THE PARTS RELATED TO THE NOTE OR DIMENSION.	
8. EXTENSION LINE		A THIN LINE USED TO EXTEND THE SIZE OF THE OBJECT BEING DRAWN SO THAT DIMENSION CAN BE PLACED NEAR THE VIEW.	A THIN LINE THAT PROJECTS FROM AN OBJECT.	
9. DIMENSION LINE		A THIN LINE WITH ARROWHEADS ON EACH END THAT TOUCH THE EXTENSION LINE. IT IS OPENED FOR THE DIMENSION.	A THIN LINE THE LENGTH OF EACH DIMENSION.	
10. LONG BREAK LINE		A THIN LINE WITH SEVERAL FREEHAND ZIG-ZAGS. IT CUTS OFF A WIDE OBJECT.	A THIN LINE WITH ZIG-ZAGS EVERY INCH OR TWO.	
11. SHORT BREAK LINE		A HEAVY WAVY LINE. IT CUTS OFF A NARROW LONG OBJECT	A WIDE LINE.	

Figure 10-5: The alphabet of lines is the drafters professional alphabet.

SELF CHECK

1. What are the different line widths in the alphabet of lines?
2. What are very wide lines used for?
3. List seven different uses for thin lines.
4. What kind of lines do not appear on a finished drawing?

Problems for Unit 10

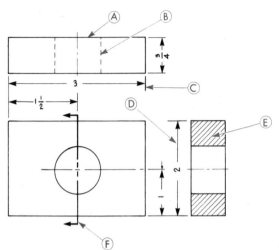

Ⓐ	_____
Ⓑ	_____
Ⓒ	_____
Ⓓ	_____
Ⓔ	_____
Ⓕ	_____

Problem 10-1: Name the lines indicated by letters on this drawing. Letter your answers.

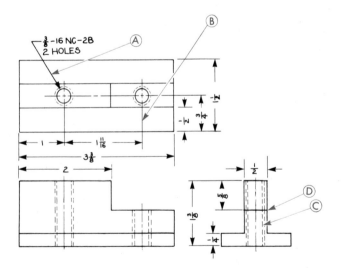

Ⓐ	_____
Ⓑ	_____
Ⓒ	_____
Ⓓ	_____

Problem 10-2: Name the lines indicated by letters on this drawing. Letter your answers.

Sheet layout is the first step a drafter must make before the mechanical drawing is started. The sheet layout consists of:

● Border lines
● Title block
● Lettering

Each classroom's sheet layout may be different. The layout depends on the size of the paper and the teacher's requirements. Most schools use two standard paper sizes. "A" size is 8-1/2 x 11 inches. "B" size is 11 x 17 inches.

1. **APPROVED** — TEACHER'S APPROVAL OF FINISHED DRAWING

2. **GRADE** — TEACHER'S GRADE FOR DRAWING

3. **DRAFTING I** — STUDENT'S DRAFTING LEVEL

4. **SAN CARLOS JUNIOR HIGH SCHOOL** — STUDENT'S SCHOOL

5. **SHAFT BRACKET** — NAME OF THE PART DRAWN.

6. **DRAWN BY** — NAME OF THE STUDENT DRAFTER

7. **SCALE** — THE SIZE OF THE DRAWING'S SCALE

8. **FEB. 24, 1980** — THE DATE OF THE COMPLETED DRAWING

9. **DR. NO.** — THE DRAWING NUMBER

Figure 11-1: Title strip information for classroom work.

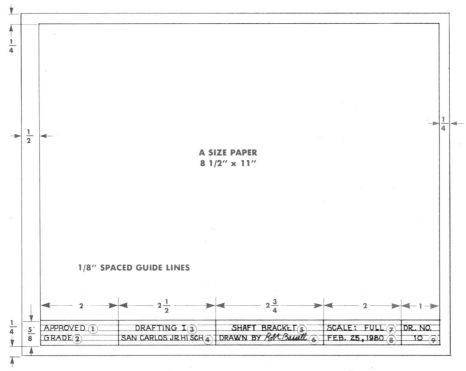

Figure 11-2: "A" size paper with a horizontal sheet layout.

Figure 11-1 shows the title strip information usually needed for a drawing. An "A" size sheet layout can be used horizontally (figure 11-2) or vertically (figure 11-3). "B" size sheet layouts are usually horizontal (figure 11-4).

Sheet layouts and title blocks for industry need more information. Each industry has its own style of sheet layout to meet its needs. Figure 11-5 shows several examples of an industrial title block.

Whether you are in school or industry, having all the drawings the same size with the same sheet layout makes handling, checking and storing easier (figure 11-6).

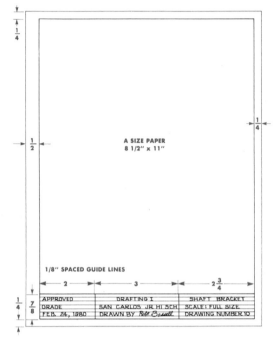

Figure 11-3: "A" size paper with a vertical sheet layout.

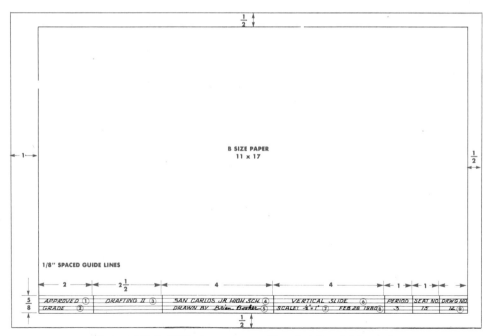

Figure 11-4: "B" size paper with a horizontal layout.

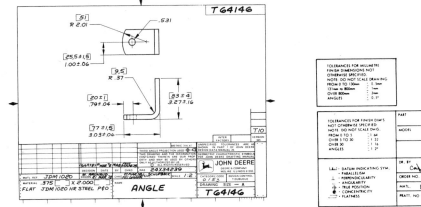

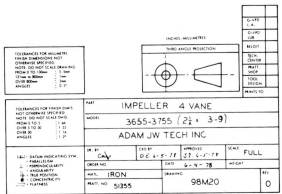

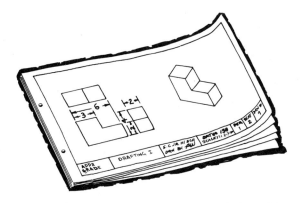

Figure 11-5: Industrial title blocks are designed for the needs of the job.

Figure 11-6: Uniform size paper and sheet layouts are easier to handle.

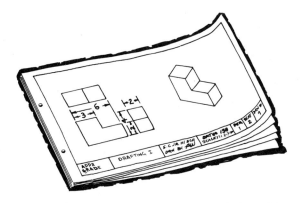

SELF CHECK

1. What is the size of "A" paper?
2. What is the size of "B" paper?
3. List the three parts of a sheet layout.
4. Why do industrial title blocks have more information than those used on school drawings?

3 BASIC DRAFTING

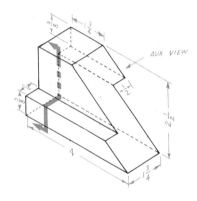

The next 14 units will teach you about basic drafting. They will prepare you to draw, to read blueprints, and to understand the basic ideas which are necessary to do the drawings needed by industry.

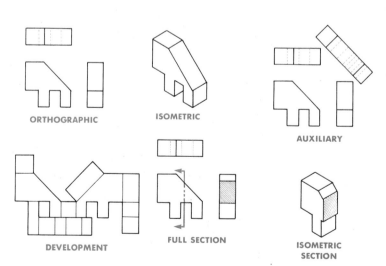

Figure III-1: The basic types of technical drawings.

Everything is made of geometric shapes. All shapes can be drawn with straight and curved lines.

The basic two-dimensional or flat geometric forms are shown in figure 12-1. They are defined in figures 12-2 through 12-4.

Basic three-dimensional forms are shown in figure 12-5. Learn as many of these basic forms as you can.

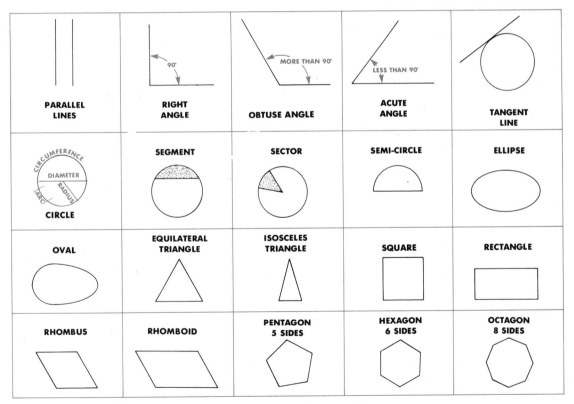

Figure 12-1: Two-dimensional forms.

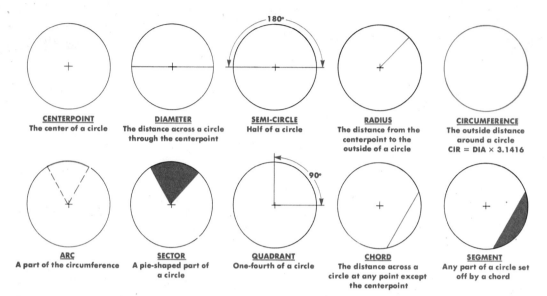

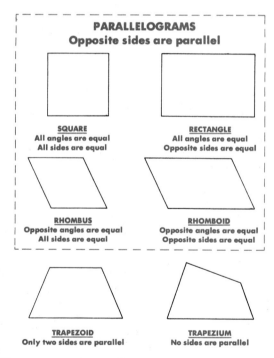

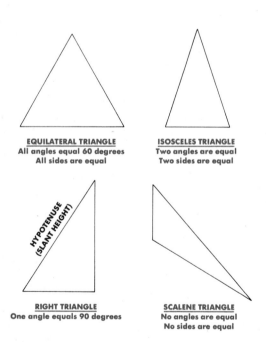

Figure 12-2: A circle is a plane (flat) curved figure. All points are the same distance from its center. All circles have 360 degrees.

Figure 12-3: A triangle is a three sided plane (flat) figure. All triangles have three sides. All triangles have 180 degrees.

Figure 12-4: A quadrilateral is a four sided plane (flat) figure. All quadrilaterals have four sides. All quadrilaterals have 360 degrees.

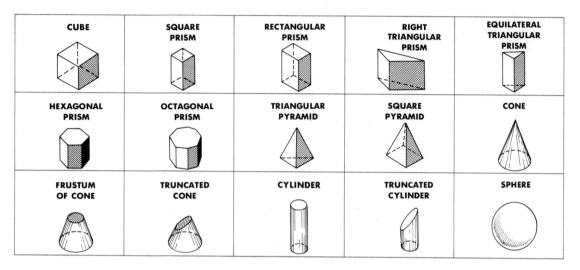

Figure 12-5: Three-dimensional forms.

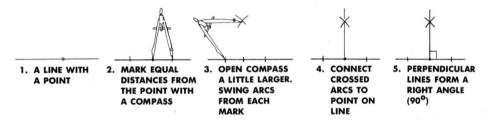

Figure 12-6: Construction of a perpendicular line from a point on a line.

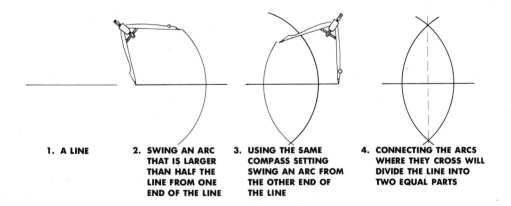

Figure 12-7: Bisecting a line.

Geometric constructions have many different types of forms and ways to draw them. Figures 12-6 through 12-18 are a few of the basic constructions that can be drawn with basic drafting tools.

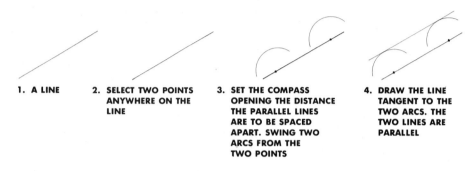

1. A LINE

2. SELECT TWO POINTS ANYWHERE ON THE LINE

3. SET THE COMPASS OPENING THE DISTANCE THE PARALLEL LINES ARE TO BE SPACED APART. SWING TWO ARCS FROM THE TWO POINTS

4. DRAW THE LINE TANGENT TO THE TWO ARCS. THE TWO LINES ARE PARALLEL

Figure 12-8: How to draw parallel lines with a compass and ruler.

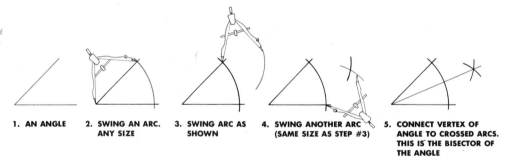

1. AN ANGLE

2. SWING AN ARC. ANY SIZE

3. SWING ARC AS SHOWN

4. SWING ANOTHER ARC (SAME SIZE AS STEP #3)

5. CONNECT VERTEX OF ANGLE TO CROSSED ARCS. THIS IS THE BISECTOR OF THE ANGLE

Figure 12-9: Bisecting an angle with a compass.

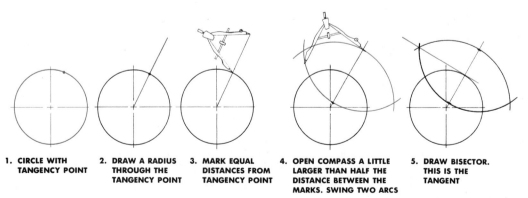

1. CIRCLE WITH TANGENCY POINT

2. DRAW A RADIUS THROUGH THE TANGENCY POINT

3. MARK EQUAL DISTANCES FROM TANGENCY POINT

4. OPEN COMPASS A LITTLE LARGER THAN HALF THE DISTANCE BETWEEN THE MARKS. SWING TWO ARCS

5. DRAW BISECTOR. THIS IS THE TANGENT

Figure 12-10: Drawing a line tangent to a point on a circle.

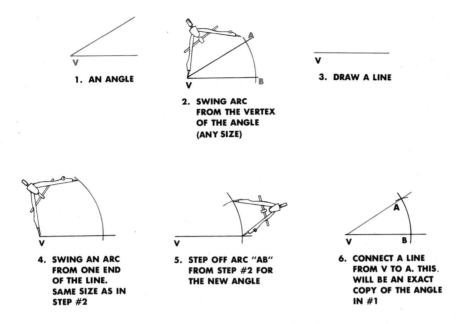

1. AN ANGLE

2. SWING ARC FROM THE VERTEX OF THE ANGLE (ANY SIZE)

3. DRAW A LINE

4. SWING AN ARC FROM ONE END OF THE LINE. SAME SIZE AS IN STEP #2

5. STEP OFF ARC "AB" FROM STEP #2 FOR THE NEW ANGLE

6. CONNECT A LINE FROM V TO A. THIS WILL BE AN EXACT COPY OF THE ANGLE IN #1

Figure 12-11: A compass can be used to copy an angle.

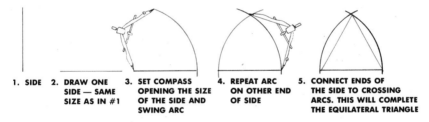

1. SIDE

2. DRAW ONE SIDE — SAME SIZE AS IN #1

3. SET COMPASS OPENING THE SIZE OF THE SIDE AND SWING ARC

4. REPEAT ARC ON OTHER END OF SIDE

5. CONNECT ENDS OF THE SIDE TO CROSSING ARCS. THIS WILL COMPLETE THE EQUILATERAL TRIANGLE

Figure 12-12: Construction of an equilateral triangle (all sides equal).

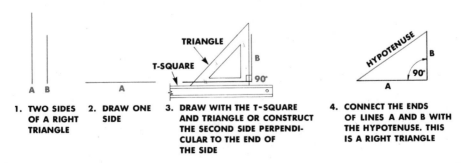

1. TWO SIDES OF A RIGHT TRIANGLE

2. DRAW ONE SIDE

3. DRAW WITH THE T-SQUARE AND TRIANGLE OR CONSTRUCT THE SECOND SIDE PERPENDICULAR TO THE END OF THE SIDE

4. CONNECT THE ENDS OF LINES A AND B WITH THE HYPOTENUSE. THIS IS A RIGHT TRIANGLE

Figure 12-13: Construct a right triangle from two sides with a T-square and triangle.

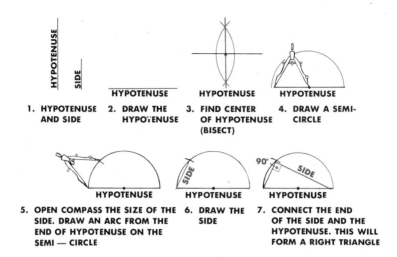

Figure 12-14: Construction of a right triangle from the hypotenuse and a side.

1. HYPOTENUSE AND SIDE
2. DRAW THE HYPOTENUSE
3. FIND CENTER OF HYPOTENUSE (BISECT)
4. DRAW A SEMI-CIRCLE
5. OPEN COMPASS THE SIZE OF THE SIDE. DRAW AN ARC FROM THE END OF HYPOTENUSE ON THE SEMI — CIRCLE
6. DRAW THE SIDE
7. CONNECT THE END OF THE SIDE AND THE HYPOTENUSE. THIS WILL FORM A RIGHT TRIANGLE

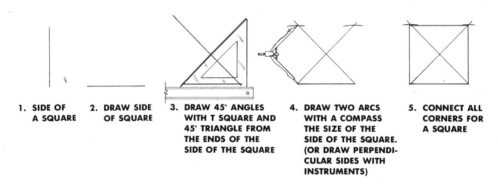

Figure 12-15: Construction of a square from a side.

1. SIDE OF A SQUARE
2. DRAW SIDE OF SQUARE
3. DRAW 45° ANGLES WITH T SQUARE AND 45° TRIANGLE FROM THE ENDS OF THE SIDE OF THE SQUARE
4. DRAW TWO ARCS WITH A COMPASS THE SIZE OF THE SIDE OF THE SQUARE. (OR DRAW PERPENDICULAR SIDES WITH INSTRUMENTS)
5. CONNECT ALL CORNERS FOR A SQUARE

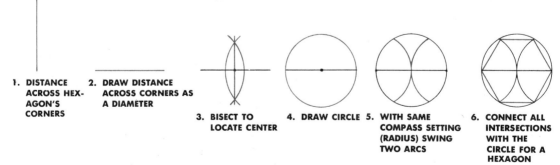

1. DISTANCE ACROSS HEXAGON'S CORNERS
2. DRAW DISTANCE ACROSS CORNERS AS A DIAMETER
3. BISECT TO LOCATE CENTER
4. DRAW CIRCLE
5. WITH SAME COMPASS SETTING (RADIUS) SWING TWO ARCS
6. CONNECT ALL INTERSECTIONS WITH THE CIRCLE FOR A HEXAGON

Figure 12-16: How to draw a hexagon with a compass and ruler.

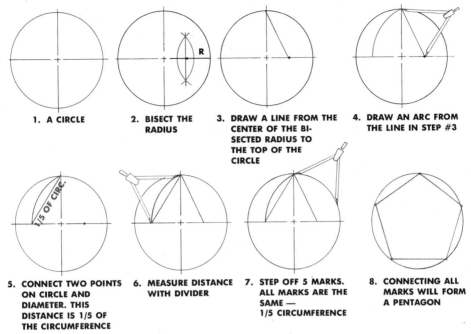

Figure 12-17: Drawing a pentagon (five sides) inside a circle.

1. A CIRCLE

2. BISECT THE RADIUS

3. DRAW A LINE FROM THE CENTER OF THE BI-SECTED RADIUS TO THE TOP OF THE CIRCLE

4. DRAW AN ARC FROM THE LINE IN STEP #3

5. CONNECT TWO POINTS ON CIRCLE AND DIAMETER. THIS DISTANCE IS 1/5 OF THE CIRCUMFERENCE

6. MEASURE DISTANCE WITH DIVIDER

7. STEP OFF 5 MARKS. ALL MARKS ARE THE SAME — 1/5 CIRCUMFERENCE

8. CONNECTING ALL MARKS WILL FORM A PENTAGON

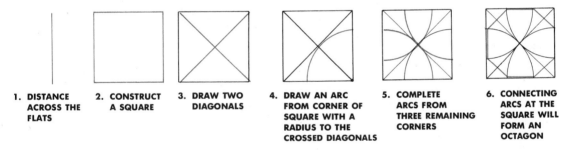

1. DISTANCE ACROSS THE FLATS

2. CONSTRUCT A SQUARE

3. DRAW TWO DIAGONALS

4. DRAW AN ARC FROM CORNER OF SQUARE WITH A RADIUS TO THE CROSSED DIAGONALS

5. COMPLETE ARCS FROM THREE REMAINING CORNERS

6. CONNECTING ARCS AT THE SQUARE WILL FORM AN OCTAGON

Figure 12-18: Drawing an octogon (eight sides) from the distance across its flats.

SELF CHECK

1. What is a right angle?
2. What is a cube?
3. List four kinds of triangles.
4. List and define the parts of a circle.

Problems for Unit 12

Problem 12-1: Draw a horizontal line of any convenient length and bisect it.

Problem 12-2: Given a 2-1/2 inch diameter circle, construct a pentagon.

Problem 12-3: Draw a four inch diameter circle. Draw, identify, and label the following parts:

a. diameter
b. radius
c. circumference
d. arc

e. chord
f. quadrant
g. sector
h. segment

Problem 12-4:
a. construct an equilateral triangle with 1-1/2 inch sides.

b. construct a right triangle 2 inches high with a 1-1/2 inch base.

c. construct a scalene triangle with sides 1-1/4, 2, and 2-1/2 inches long.

Problem 12-5: Copy angle A B C in a new location.

Problem 12-6: Bisect angle A B C in problem 5.

Problem 12-7: Construct a line 3/4 inch from, and parallel to, line B C in Problem 5.

Problem 12-8: Construct a square with 1-1/2 inch sides.

Problem 12-9: Draw a line tangent to point C.

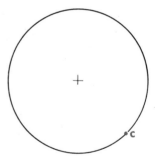

Problem 12-10: Construct a 1-1/2 inch hexagon.

Orthographic drawings, also called <u>multiview</u> or <u>working drawings</u>, are the most used drawings in industry. In some orthographic drawings, a single view of a simple object gives enough information for a worker to build the object (figure 13-1). However, most orthographic drawings require <u>two views</u> (figure 13-2) or <u>three views</u> (figure 13-3).

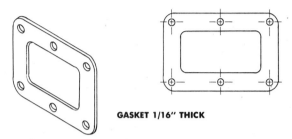

GASKET 1/16" THICK

Figure 13-1: One-view drawings are adequate for simple objects.

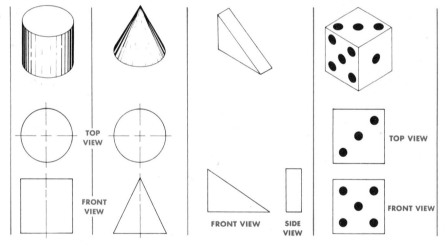

Figure 13-2: Three-dimensional objects usually need two or more views.

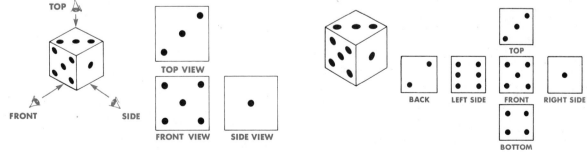

Figure 13-3: A three-view working drawing.

Figure 13-4: A six-view drawing.

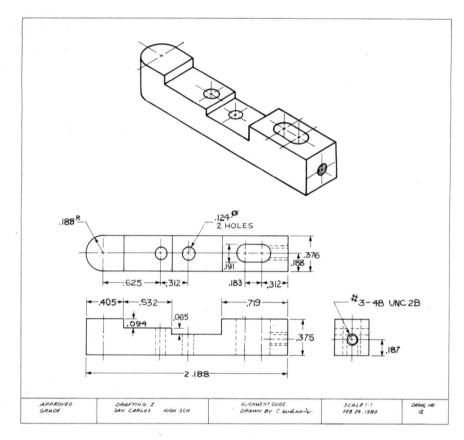

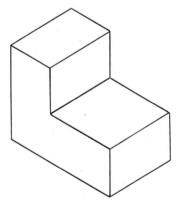

Figure 13-5: A working drawing drawn by a high school student.

If necessary, an orthographic drawing can consist of as many as six views (figure 13-4). An example of a working drawing that would be used in industry is shown in figure 13-5.

Orthographic projection starts with the <u>front view.</u> This view is usually the largest view of the drawing. It shows the shape of the object in the most detail. On the drawing, the front view appears as if you were looking at the object straight on (figure 13-6).

Figure 13-6: Front views are usually the most descriptive view.

Orthographic projection continues with the top view. The top view (shown as if the top is turned toward you) is projected straight up from the front view (figure 13-7). Leave enough space between the views for dimensions.

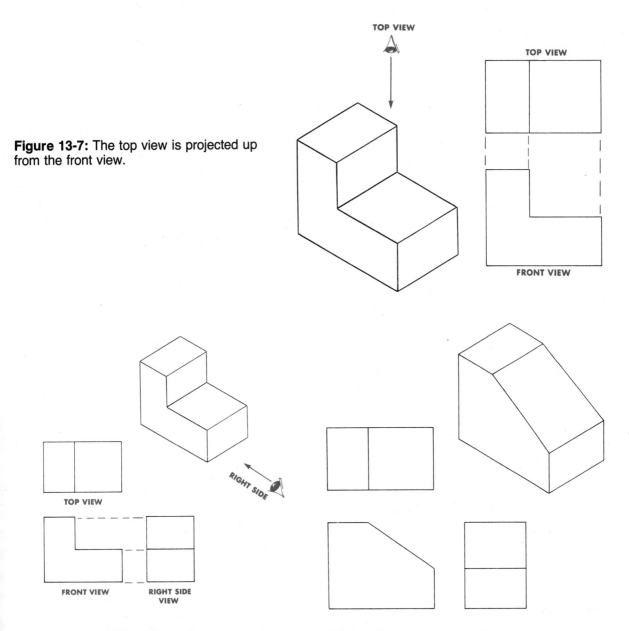

Figure 13-7: The top view is projected up from the front view.

Figure 13-8: Projection of the right side view from the front view.

Figure 13-9: The depth of the front view and side view are the same.

After the front view and the top view are drawn, the side view is projected out to the right side of the front view (figure 13-8). The depth of the top view and the side view are always the same (figure 13-9).

When any two views are drawn, the third view can be fully projected as shown in figures 13-10 and 13-11.

Follow the alphabet of lines as explained in Unit 10. An example of the line work in a multiview drawing is shown in figure 13-12.

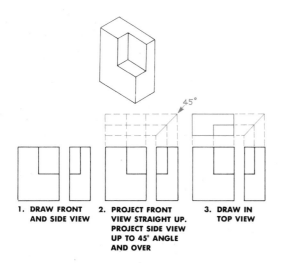

1. DRAW FRONT AND SIDE VIEW
2. PROJECT FRONT VIEW STRAIGHT UP. PROJECT SIDE VIEW UP TO 45° ANGLE AND OVER
3. DRAW IN TOP VIEW

Figure 13-10: Projecting the top view from the front and side views.

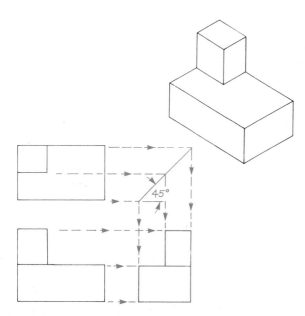

Figure 13-11: Projecting the side view from the front and top views.

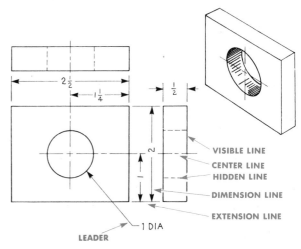

VISIBLE LINE
CENTER LINE
HIDDEN LINE
DIMENSION LINE
EXTENSION LINE
1 DIA
LEADER

Figure 13-12: The alphabet of lines on an orthographic drawing.

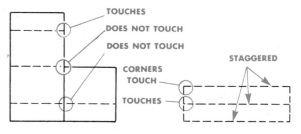

TOUCHES
DOES NOT TOUCH
DOES NOT TOUCH
CORNERS TOUCH
TOUCHES
STAGGERED

Figure 13-13: Rules for hidden lines include staggering the dashes in parallel lines and not touching intersecting lines.

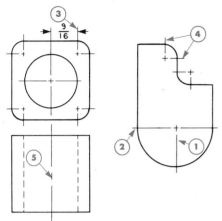

1. TWO TIMES THINNER THAN OBJECT LINES
2. PASSES THROUGH TANGENCIES AND LINES ABOUT 1/8"
3. CAN BE USED FOR EXTENSION LINES
4. TWO CENTER LINES USED TO LOCATE CENTER OF CIRCLE
5. ONE CENTER LINE IS USED ON SIDE VIEW OF HOLE

Figure 13-14: Rules for center lines.

The rules for hidden lines are shown in figure 13-13.

Center line rules are shown in figure 13-14.

SELF CHECK

1. Which type of mechanical drawing is used most often?
2. Name the three main views used in orthographic drawings.
3. How many views can a drawing have?
4. How is the front view chosen?

Problems for Unit 13

Problem 13-1: Draw the required orthographic views as assigned by your instructor.

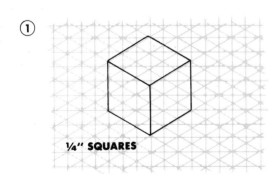

① ¼" SQUARES

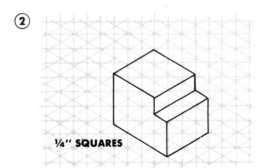

② ¼" SQUARES

③

¼″ **SQUARES**

④

¼″ **SQUARES**

⑤

¼″ **SQUARES**

⑥

¼″ **SQUARES**

⑦

¼″ **SQUARES**

⑧

¼″ **SQUARES**

⑨

¼″ **SQUARES**

⑩

¼″ **SQUARES**

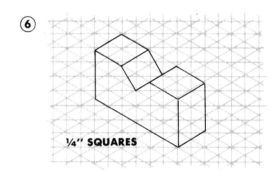
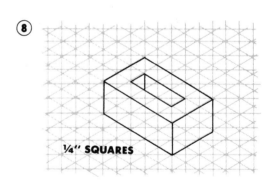
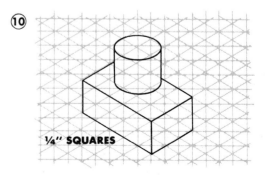

11

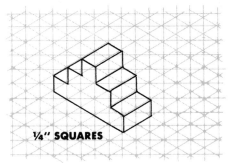

¼" SQUARES

12

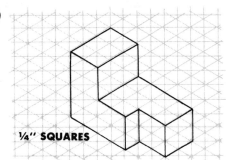

¼" SQUARES

13

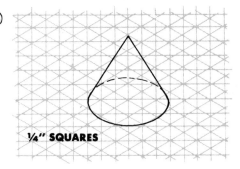

¼" SQUARES

14

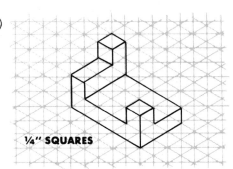

¼" SQUARES

15

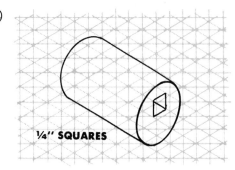

¼" SQUARES

16

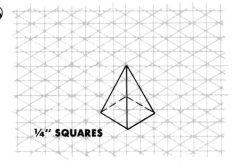

¼" SQUARES

17

¼" SQUARES

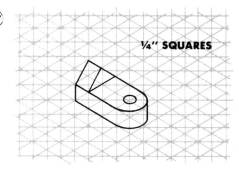

18

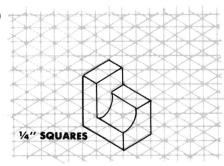

¼" SQUARES

Problem 13-2: Complete the missing views as assigned by your instructor.

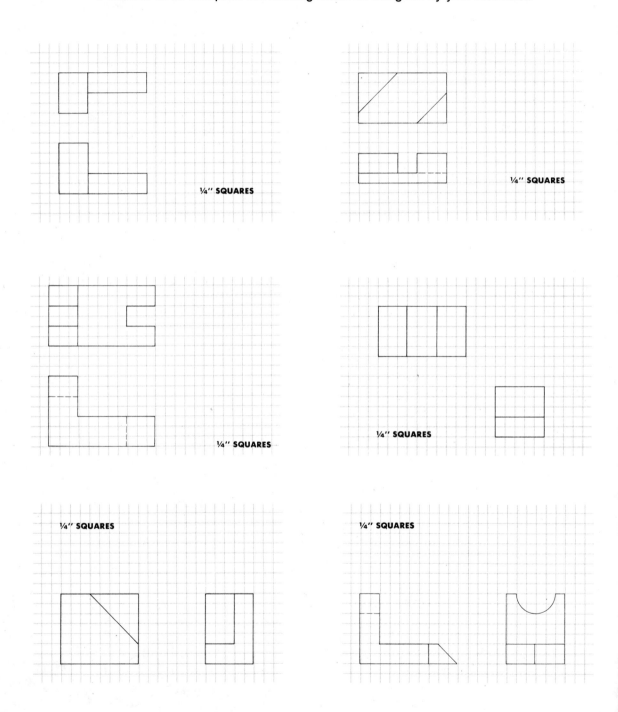

Dimensioning a drawing means to show sizes on the drawing. It is possible to draw an object without giving sizes, but can you build the object from such a drawing (figure 14-1)? To build an object from a drawing, you need its size or dimensions (figure 14-2).

Dimensioning standards are the same all over the world. Mechanical drawings from any country in the

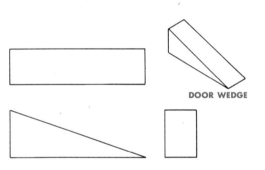

Figure 14-1: Can you make this door wedge from this drawing?

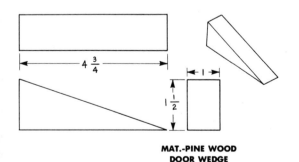

Figure 14-2: Why is it easier to make this door wedge from this drawing?

DIMENSION LINE

Figure 14-3: A dimension line is the exact length of the dimension with an opening in the center for the numerical note.

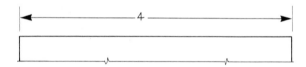

Figure 14-4: Extension lines come out at a 90° angle from the drawing. They do not touch the drawing. The gap should be about 1/16 inch.

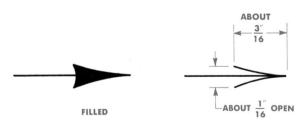

FILLED

ABOUT

Figure 14-5: Arrowheads are drawn three times longer than their width. They should touch the ends of the dimension line.

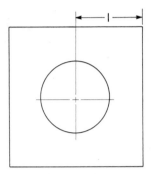

Figure 14-6: A center line can be used as an extension line.

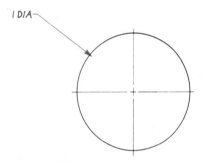

Figure 14-7: Leaders are special dimension lines used to dimension circular size. They start with a short, flat line. It bends approximately 45° to touch the specific part. The arrow should point to the center.

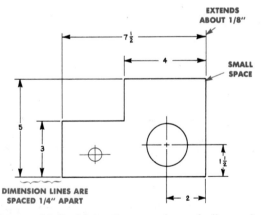

Figure 14-8: Note the spacing of dimension lines.

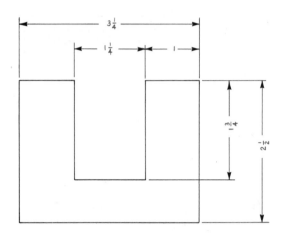

Figure 14-9: Aligned side dimensions are read from the right side. Other dimensions are read straight up and down.

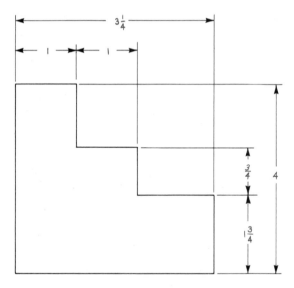

Figure 14-10: Unidirectional dimensions are all read straight up and down. Do not mix aligned and unidirectional systems.

world will look like drawings from another country because they all follow the same rules.

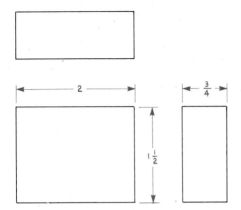

Figure 14-11: The rectangular prism is a basic form with length, width and height. Place dimensions between drawings when possible.

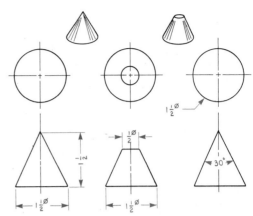

Figure 14-13: Several different ways to dimension a cone shape.

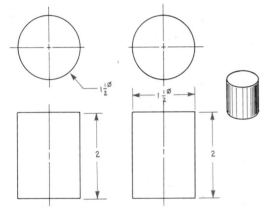

Figure 14-12: Cylinders can be dimensioned either way.

Figure 14-14: Different methods of dimensioning a pyramid form.

Dimensioning symbols have the same meaning in every drawing. Look at the dimensioning symbols on these pages. As you use them you will memorize their meanings.

Dimensioning styles vary more than symbols. Styles are concerned with where and how symbols are used in different situations. Be sure to note the

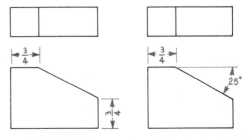

Figure 14-15: There are two techniques used to dimension an inclined surface.

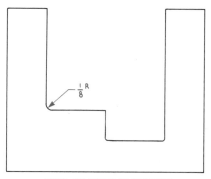

NOTE: ALL FILLETS 1/16" RADIUS
UNLESS SPECIFIED.

Figure 14-16: Fillets are inside curves. They are dimensioned or explained with a note on the drawing

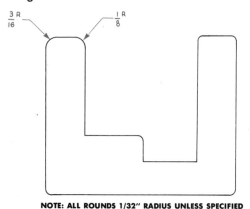

NOTE: ALL ROUNDS 1/32" RADIUS UNLESS SPECIFIED

Figure 14-17: A round is an outside curve. Rounds can be dimensioned or explained with a note on the drawing.

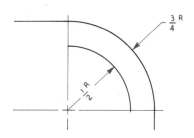

Figure 14-19: Two methods of dimensioning a radius.

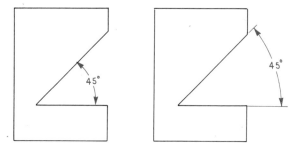

Figure 14-20: Methods for dimensioning an angle.

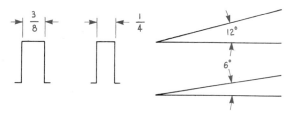

Figure 14-21: When dimensioning a small area, you can draw the arrowheads and the dimension outside the extension lines.

illustrations of the different dimensioning styles in figures 14-3 through 14-21.

Special tool operations usually need dimensions and a note (figures 14-22

Figure 14-18: Methods used to dimension the circle view. "D" or ϕ are optional symbols for diameter.

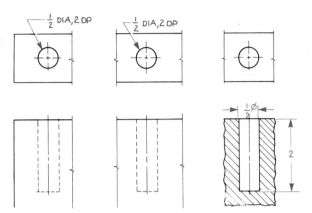

Figure 14-22: Two dimensions are needed to describe a hole: diameter and depth.

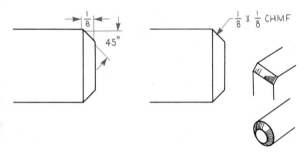

Figure 14-23: Alternate methods to dimension chamfers.

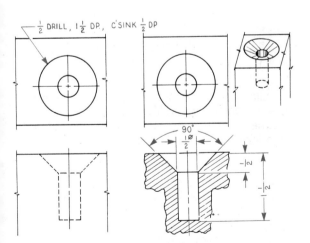

Figure 14-24: Dimensioning a countersink.

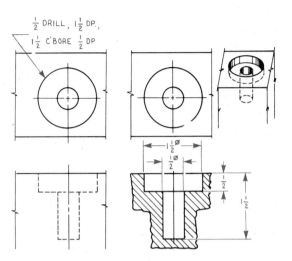

Figure 14-25: Alternate methods to dimension a counterbored hole.

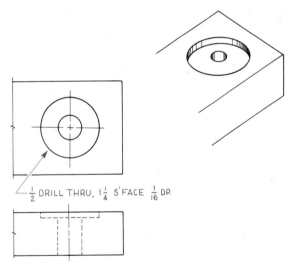

Figure 14-26: Dimensioning a spot-faced hole.

through 14-28). A completely dimensioned drawing is shown in figure 14-29.

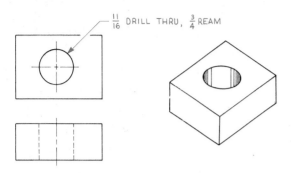

Figure 14-27: Dimensioning a drilled and reamed hole. The ream smooths the sides of a drilled hole.

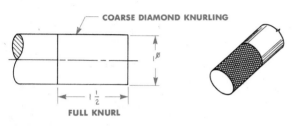

Figure 14-28: Dimensioning knurls.

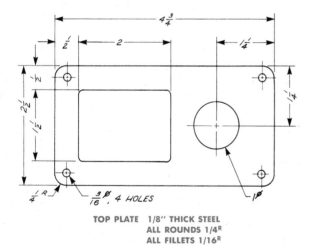

Figure 14-29: A dimensioned drawing.

Problems for Unit 14

As assigned by your instructor, use a mechanical scale, an engineer's scale, and/or a metric scale to dimension the orthographic drawings you made for Unit 13.

An isometric drawing is one type of pictorial·drawing (figure 15-1). In mechanical drafting, isometric drawings are the most often used type of pictorial drawing. Isometric drawings are fast to draw, and they actually look like the object being drawn (figure 15-2). Iso-

metrics may also be used in exploded views to show how things may be put together (figure 15-3).

In an isometric drawing, most straight lines are drawn vertically or on a 30° angle. They are drawn at their true length (figure 15-4), and can be

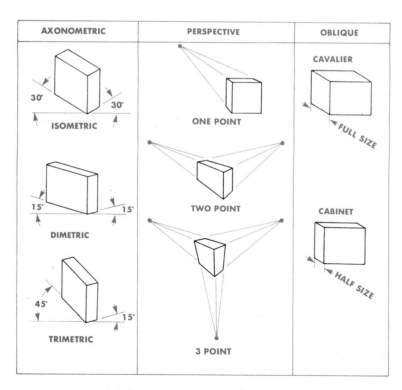

Figure 15-1: Types of pictorial drawings.

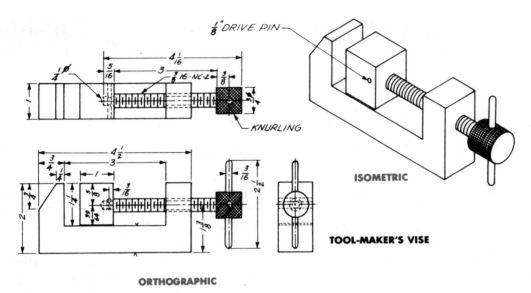

Figure 15-2: An example of a combined isometric and orthographic working drawing.

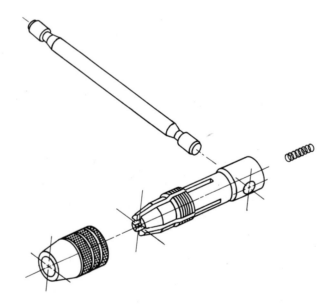

Figure 15-3: An exploded isometric drawing of a tap wrench.

measured. The steps required to draw an isometric cube are shown in figure 15-5.

Any line on an isometric drawing that is not drawn vertically or on a 30° angle is called a non-isometric line. A non-isometric line is not true length. Non-isometric lines and angles cannot be measured on the drawing. To draw non-isometric lines and angles, mark

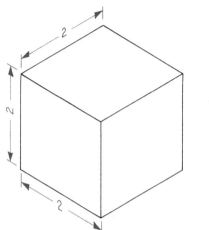

Figure 15-4: All vertical and all 30° lines in an isometric drawing are true size.

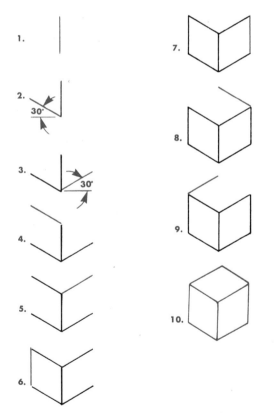

Figure 15-5: Steps for drawing an isometric cube. All lines are true length. All lines are vertical and 30°.

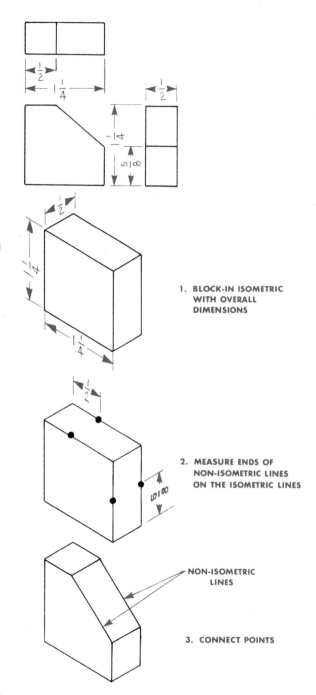

1. **BLOCK-IN ISOMETRIC WITH OVERALL DIMENSIONS**

2. **MEASURE ENDS OF NON-ISOMETRIC LINES ON THE ISOMETRIC LINES**

NON-ISOMETRIC LINES

3. **CONNECT POINTS**

Figure 15-6: Drawing non-isometric lines.

the end of the non-isometric lines on the true length lines as shown in figure 15-6.

Follow the steps in figure 15-7 to draw a non-isometric surface as an oblique surface.

A circle will appear as an ellipse on an isometric drawing (figure 15-8). The easiest way to draw a circle on an isometric drawing is with an isometric

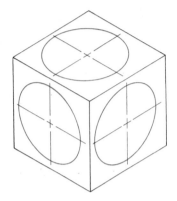

Figure 15-8: Isometric circles.

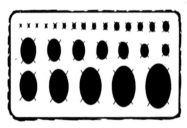

Figure 15-9: An isometric circle template.

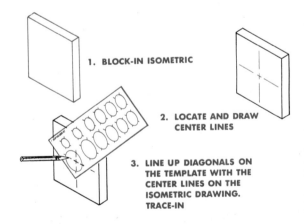

1. **BLOCK-IN ISOMETRIC**

2. **LOCATE AND DRAW CENTER LINES**

3. **LINE UP DIAGONALS ON THE TEMPLATE WITH THE CENTER LINES ON THE ISOMETRIC DRAWING. TRACE-IN**

1. **BLOCK-IN ISOMETRIC**

2. **MEASURE CORNERS OF OBLIQUE SURFACE ON ISOMETRIC LINES**

3. **CONNECT CORNERS**

Figure 15-7: Drawing a non-isometric surface.

Figure 15-10: Steps to draw an isometric circle.

circle template (figure 15-9). Follow the steps in figure 15-10, and be very careful where you place the template on the isometric center lines. Isometric center lines are explained in figure 15-11.

Hidden lines are not used on an isometric drawing unless they are necessary to show details that could not be seen otherwise (figure 15-12).

When a fully-dimensioned orthographic drawing accompanies an iso-

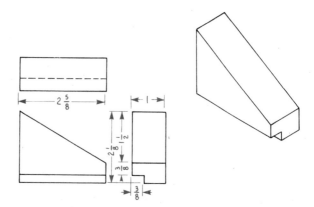

Figure 15-13: Do not dimension an isometric drawing if there is a dimensioned orthographic drawing.

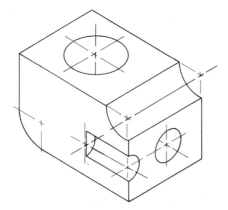

Figure 15-11: Isometric center lines are needed to align the isometric circle template.

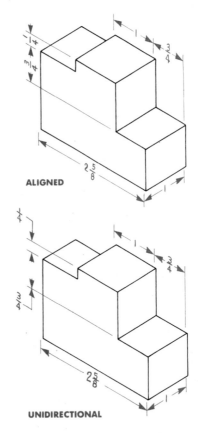

ALIGNED

UNIDIRECTIONAL

Figure 15-14: Isometric dimensioning systems.

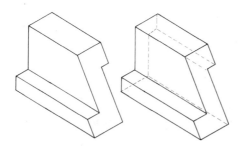

Figure 15-12: Hidden lines show how the other side of an isometric looks.

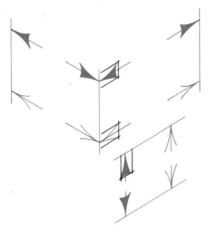

Figure 15-15: Isometric arrowheads. Practice their forms, keeping the back of the arrowhead parallel to the extension line.

metric view (figure 15-13), it is not necessary to dimension it. However, an isometric drawn alone must be dimen-

sioned with aligned or uni-directional dimensions as shown in figure 15-14. Isometric arrowheads are drawn as shown in figure 15-15.

SELF CHECK

1. What is an isometric drawing?
2. Which isometric lines are true length?
3. Which tool is used to draw an isometric circle?
4. When is it necessary to show hidden lines on an isometric drawing?

Problems for Unit 15

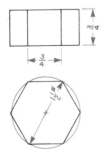

Problem 15-1: Draw the isometric and/or orthographic view as assigned by your instructor.

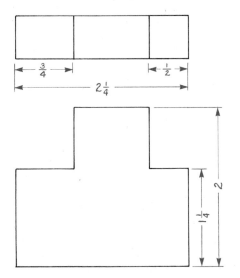

Problem 15-2: Draw the isometric and/or ortho-graphic view as assigned by your instructor.

Problem 15-3: Draw the isometric and/or ortho-graphic view as assigned by your instructor.

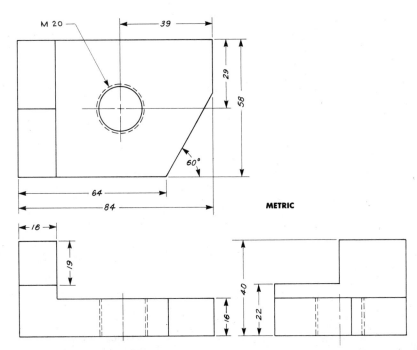

Problem 15-4: Draw the isometric and/or orthographic view as assigned by your instructor.

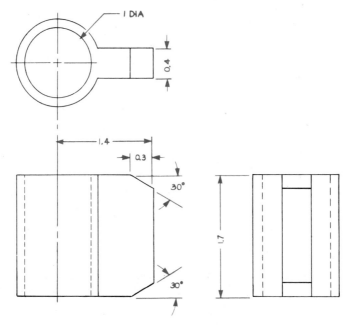

Problem 15-5: Draw the isometric and/or orthographic view as assigned by your instructor.

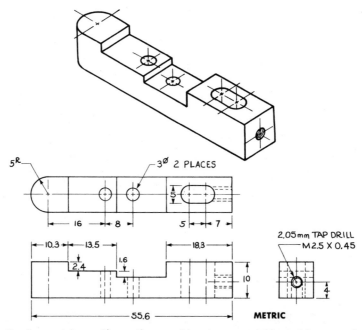

Problem 15-6: Draw the isometric and/or orthogrpahic view as assigned by your instructor.

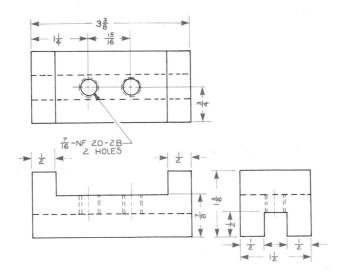

Problem 15-7: Draw the isometric and/or orthogrpahic view as assigned by your instructor.

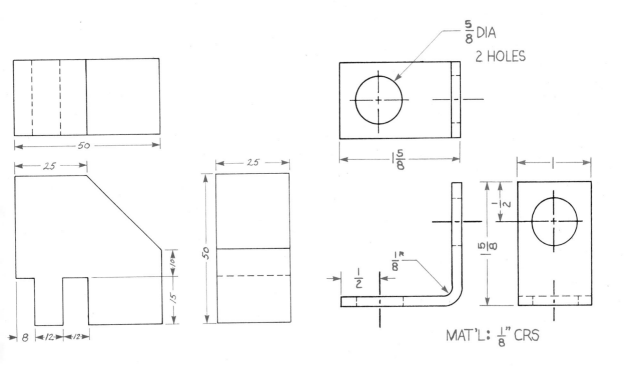

Problem 15-8: Draw the isometric and/or orthographic view as assigned by your instructor.

Problem 15-9: Draw the isometric and/or orthographic view as assigned by your instructor.

OBLIQUE DRAWING—CAVALIER

Units 16 and 17 are about oblique drawings. Oblique drawings are pictorial drawings with the face drawn the same way as the front view of an orthographic drawing.

To make an oblique drawing, project the corners back at a convenient angle (usually 30° or 45°) to show the side and top (figure 16-1). The oblique drawing is the quickest type of pictorial drawing to draw. It has two forms:
- Cavalier
- Cabinet

To make a cavalier drawing, follow the steps in figure 16-2. Note that all dimensions are full size. The advantage of a cavalier over an isometric drawing is that it is easier to draw (figure 16-3).

Cavalier drawings do have a disadvantage. The depth (distance from front to back) is full size. This makes

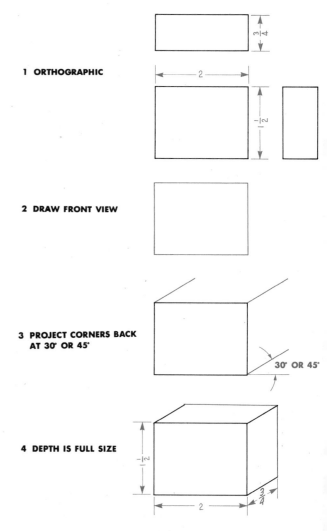

1 ORTHOGRAPHIC

2 DRAW FRONT VIEW

3 PROJECT CORNERS BACK AT 30° OR 45°

30° OR 45°

4 DEPTH IS FULL SIZE

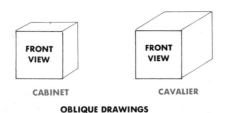

CABINET   CAVALIER

OBLIQUE DRAWINGS

Figure 16-1: Cavalier and cabinet drawings. Note the optical illusion.

Figure 16-2: Steps required to draw a cavalier drawing.

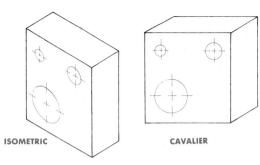

Figure 16-3: A cavalier is simpler to draw than an isometric drawing.

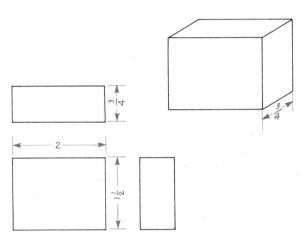

Figure 16-4: Do the depths of the orthographic and cavalier drawings look the same?

alier drawing and the orthographic drawing in figure 16-4. Which depth looks larger?

SELF CHECK

1. Name two types of oblique mechanical drawings.
2. Why are cavalier drawings used instead of isometric drawings?
3. What is the disadvantage of a cavalier drawing?
4. Which view is shared by both cavalier and orthographic drawings?

a cavalier drawing look larger than it actually is. The larger look is an optical illusion. Compare the depth of the cav-

Problems for Unit 16

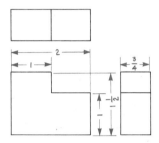

Problem 16-1: Make a cavalier and/or orthographic drawing of the object as assigned by your instructor.

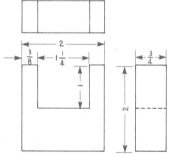

Problem 16-2: Make a cavalier and/or orthographic drawing of the object as assigned by your instructor.

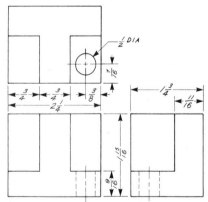

Problem 16-3: Make a cavalier and/or orthographic drawing of the object as assigned by your instructor.

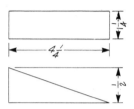

Problem 16-4: Make a cavalier and/or orthographic drawing of the object as assigned by your instructor.

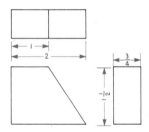

Problem 16-5: Make a cavalier and/or orthographic drawing of the object as assigned by your instructor.

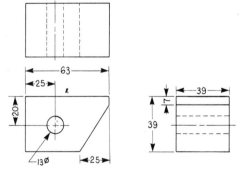

Problem 16-6: Make a cavalier and/or orthographic drawing of the object as assigned by your instructor.

The second type of oblique drawing is called a <u>cabinet drawing</u>. Cabinet drawings are used more often than cavalier drawings because the cabinet

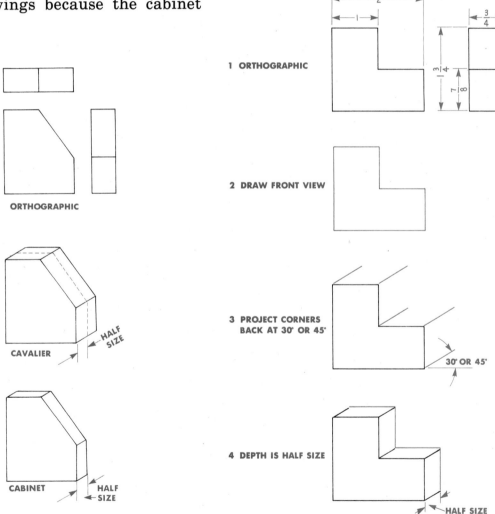

ORTHOGRAPHIC

CAVALIER HALF SIZE

CABINET HALF SIZE

Figure 17-1: A cabinet drawing looks closer to the actual form than does a cavalier drawing.

1 ORTHOGRAPHIC

2 DRAW FRONT VIEW

3 PROJECT CORNERS BACK AT 30° OR 45°

30° OR 45°

4 DEPTH IS HALF SIZE

HALF SIZE

Figure 17-2: Steps for drawing a cabinet drawing.

drawing looks more like the object than does the cavalier. To make the cabinet drawing look more realistic the depth of the object is halved (figure 17-1).

To make a cabinet drawing, follow the steps in figure 17-2. The only difference between a cabinet and a cavalier drawing is the depth. Everything else is the same.

SELF CHECK

1. Which oblique drawing is used most often by drafters?
2. Which oblique drawing has all true dimensions?
3. What is the advantage of a cavalier drawing over an isometric drawing?
4. Why are cabinet drawings made?

Problems for Unit 17

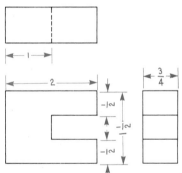

Problem 17-1: Make a cabinet and/or orthographic drawing of the object as assigned by your instructor.

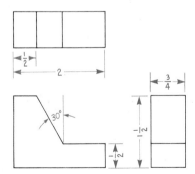

Problem 17-2: Make a cabinet and/or orthographic drawing of the object as assigned by your instructor.

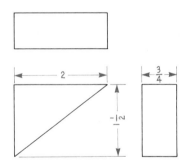

Problem 17-3: Make a cabinet and/or orthographic drawing of the object as assigned by your instructor.

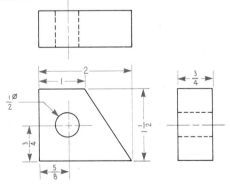

Problem 17-4: Make a cabinet and/or orthographic drawing of the object as assigned by your instructor.

One-point perspective drawing is drawn to look life-like (figure 18-1). The perspective drawing shows what an object will look like in real life and is very helpful for people who do not understand working drawings. Although this unit is concerned with one-point perspective drawings, there are several other types of perspective drawings (figure 18-2).

Figure 18-1: A life-like one-point perspective.

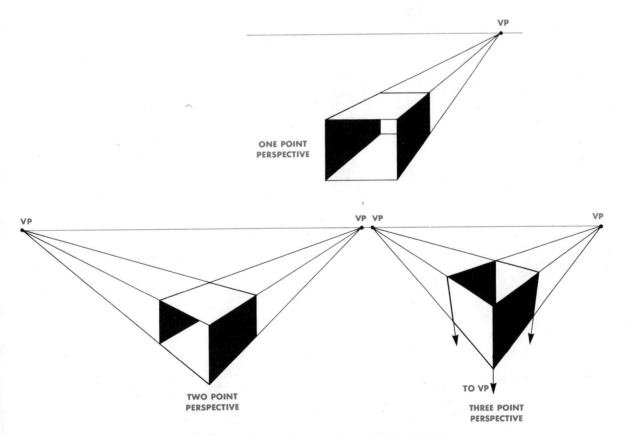

ONE POINT PERSPECTIVE

TWO POINT PERSPECTIVE

THREE POINT PERSPECTIVE

Figure 18-2: Different types of perspective drawings.

One-point perspective drawings have three main parts (figure 18-3):
● Vanishing point (VP)
● Horizon
● Object being drawn

The <u>horizon</u> in a one-point perspective drawing is always at the viewer's eye level. As the object is moved, its form changes (figure 18-4). When the object is above the horizon, you can see the object's bottom. When it is below the horizon, you can see its top. The lines from the object to the vanishing point always get smaller. This is the way objects appear in real life (figure 18-5).

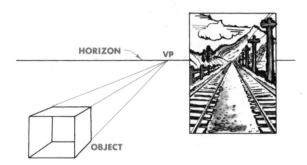

Figure 18-3: Parts of a one-point perspective.

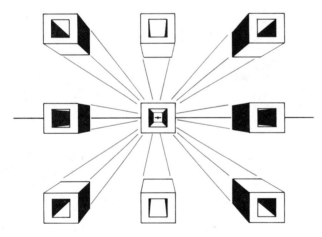

Figure 18-4: The form of an object is changed as it is moved.

Figure 18-5: A realistic one-point perspective.

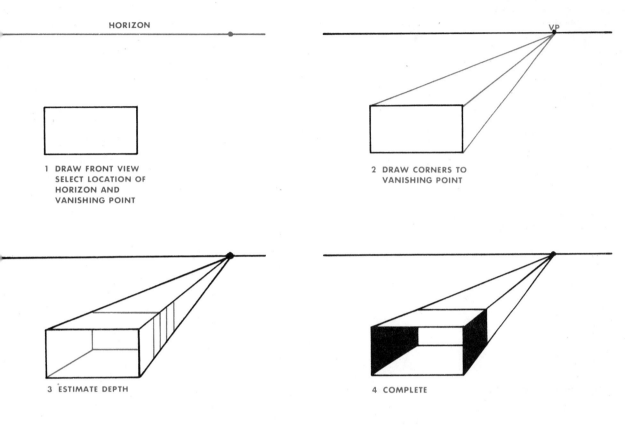

Figure 18-6: Four steps to draw a one-point perspective.

Drawing steps for making a one-point perspective are shown in figure 18-6. Do not make any measurements. Guess at the line sizes until your object looks about the right shape. These are the first steps in drawing any form in perspective.

SELF CHECK

1. What is the advantage of a perspective drawing over an isometric drawing?
2. What is at eye level on a perspective drawing?
3. What parts of a cube will be seen if it is below the horizon?
4. What other kind of drawing is the front view of a one-point perspective drawing similar to?

Problems for Unit 18

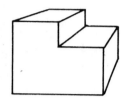

Problem 18-1: Make a one-point perspective drawing of the object.

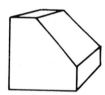

Problem 18-2: Make a one-point perspective drawing of the object.

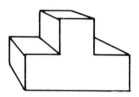

Problem 18-3: Make a one-point perspective drawing of the object.

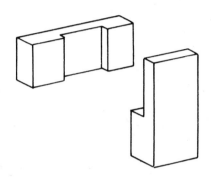

Problem 18-4: Make a one-point perspective drawing of the object.

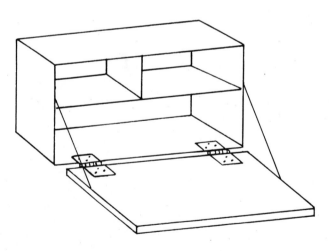

Problem 18-5: Make a one-point perspective drawing of the object.

TWO-POINT PERSPECTIVE

Two-point perspective makes pictorial drawings that are more life-like than one-point perspective drawings (figure 19-1). The main parts of two-point perspective drawings are (figure 19-2):

● Two vanishing points
● Horizon
● Object being drawn

As in one-point perspective, an object drawn to two-point perspective will look different as its position changes

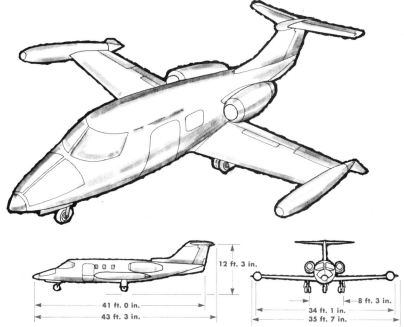

12 ft. 3 in.

41 ft. 0 in.
43 ft. 3 in.

8 ft. 3 in.
34 ft. 1 in.
35 ft. 7 in.

Figure 19-1: A two-view and a two-point perspective of a Lear jet.

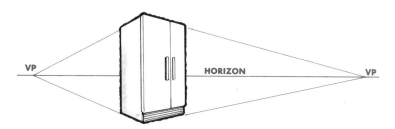

VP   HORIZON   VP

Figure 19-2: The parts of a two-point perspective.

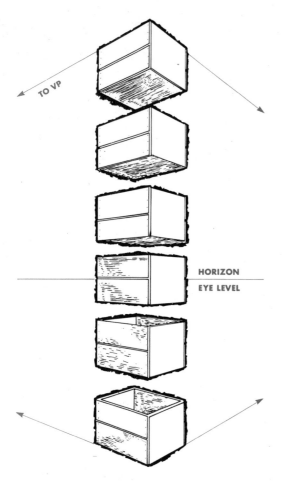

Figure 19-3: Note the visible parts of the box in relation to the horizon line.

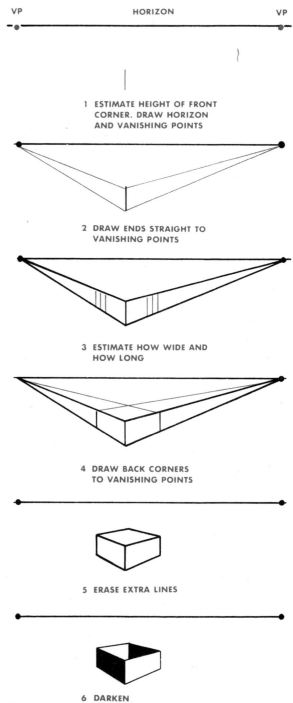

1 ESTIMATE HEIGHT OF FRONT CORNER. DRAW HORIZON AND VANISHING POINTS

2 DRAW ENDS STRAIGHT TO VANISHING POINTS

3 ESTIMATE HOW WIDE AND HOW LONG

4 DRAW BACK CORNERS TO VANISHING POINTS

5 ERASE EXTRA LINES

6 DARKEN

Figure 19-4: Steps for drawing a two-point perspective.

(figure 19-3). All views below the horizon show the top and sides. Above the horizon, all views show the bottom and sides.

The drawing steps for a two-point perspective are shown in figure 19-4. Without measurements, guess at line sizes until the object looks about right. Placing vanishing points too close together, or the object too far below the

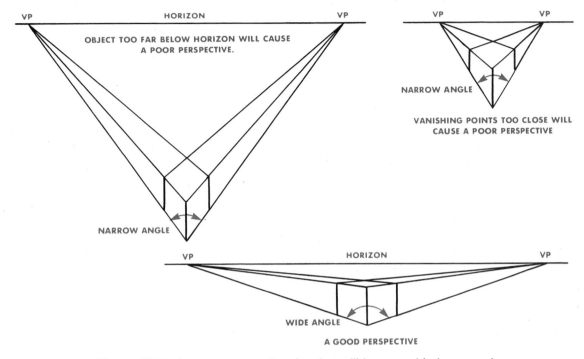

Figure 19-5: A good perspective drawing will have a wide base angle.

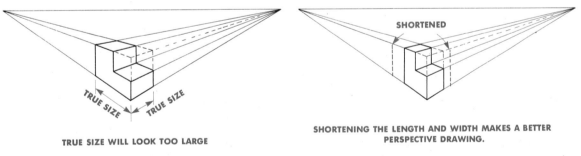

Figure 19-6: True sizes are used on orthographic and isometric drawings.

horizon will cause a poor perspective drawing (figure 19-5).

True sizes are used on orthographic and isometric drawings. Perspective drawings are shortened by the drafter as the sides near the vanishing point and get smaller in size (figure 19-6).

Perspective drawing time is speeded up if pins are placed in the vanishing points as shown in figure 19-7.

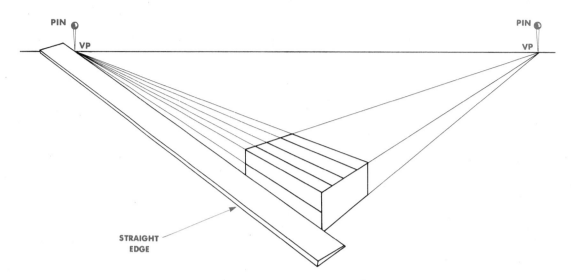

Figure 19-7: A pin in the vanishing point aids perspective drawing.

1. What is the main difference between one- and two-point perspective?
2. What happens if vanishing points are too close together?
3. Which part of a cube will be seen if the cube is on the horizon?
4. Where are the vanishing points located?

Problems for Unit 19

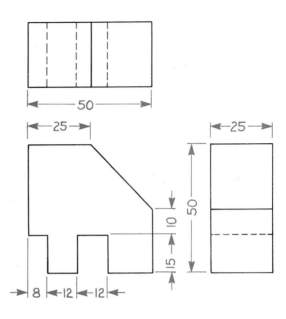

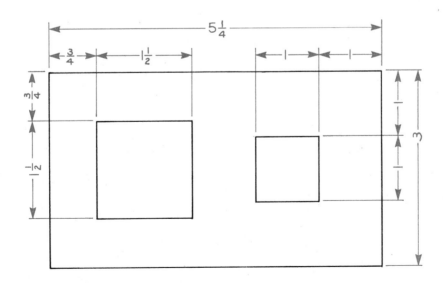

Problem 19-1: Make a two-point perspective drawing of the object.

Problem 19-3: Make a two point-perspective drawing of the object.

Problem 19-2: Make a two-point perspective drawing of the object.

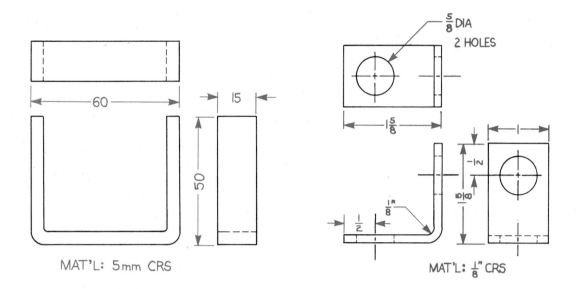

MAT'L: 5mm CRS

$\frac{5}{8}$ DIA
2 HOLES

MAT'L: $\frac{1}{8}$" CRS

Problem 19-4: Make a two-point perspective drawing of the object.

Problem 19-6: Make a two-point perspective drawing of the object.

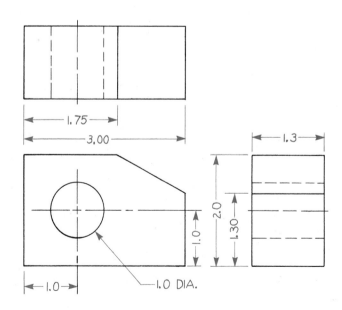

Problem 19-5: Make a two-point perspective drawing of the object.

Units 20 and 21 are about sectional drawings. A sectional drawing (figure 20-1) shows the inside of an object when it is complicated and too many hidden lines would be needed to show all the detail. A <u>sectioned view</u> (figure 20-2) will make a complicated drawing much easier to understand. The purpose of a sectional view is to show only the detail of the object that is visible along a cutting plane.

The <u>cutting plane line</u> is a heavy line drawn with a long dash followed by two short dashes and another long dash

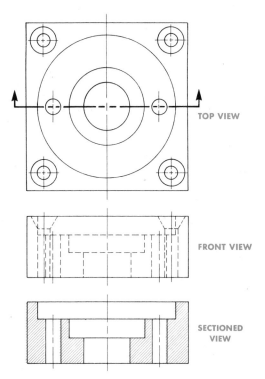

TOP VIEW

FRONT VIEW

SECTIONED VIEW

Figure 20-2: A sectioned drawing simplifies a complicated drawing.

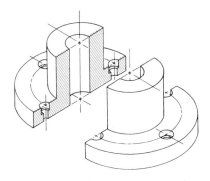

Figure 20-1: A sectional drawing shows the inside of an object.

(figure 20-3). It is drawn where the object would have to be cut open to show the inside detail. The arrowheads on the ends of a cutting plane line point

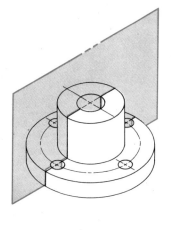

Figure 20-3: The cutting plane line shows where the section is to be located.

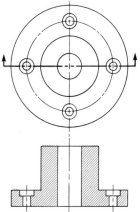

Figure 20-4: The arrowheads on the cutting plane line point in the direction in which you would have to be looking to see the sectional view.

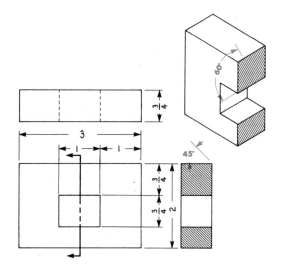

Figure 20-5: Section lines are drawn on the solid material only.

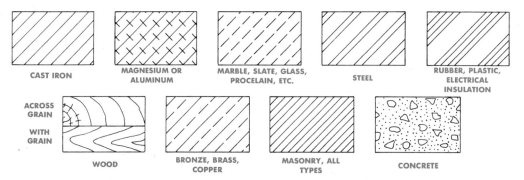

CAST IRON

MAGNESIUM OR ALUMINUM

MARBLE, SLATE, GLASS, PROCELAIN, ETC.

STEEL

RUBBER, PLASTIC, ELECTRICAL INSULATION

ACROSS GRAIN

WITH GRAIN

WOOD

BRONZE, BRASS, COPPER

MASONRY, ALL TYPES

CONCRETE

Figure 20-6: Examples of common section line symbols.

in the direction in which you would have to look to see the section view after the object was cut (figure 20-4).

Section lines (also known as cross-hatching) are thin, sharp, parallel lines that are usually drawn at a 45 degree angle on an orthographic view and at 60 degrees on an isometric view. They show the solid material that would actually be cut if the object was cut along the cutting plane line (figure 20-5).

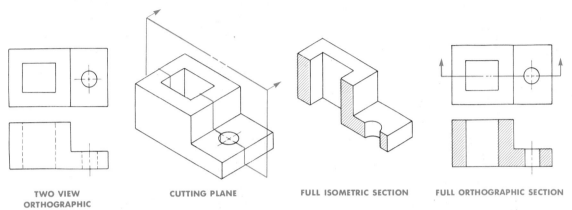

TWO VIEW ORTHOGRAPHIC CUTTING PLANE FULL ISOMETRIC SECTION FULL ORTHOGRAPHIC SECTION

Figure 20-7: A full section drawing with the cutting plane line through the top view.

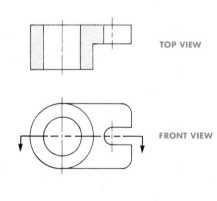

TOP VIEW

FRONT VIEW

Figure 20-8: A full section drawing with the cutting plane line through the front view.

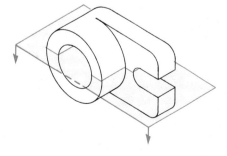

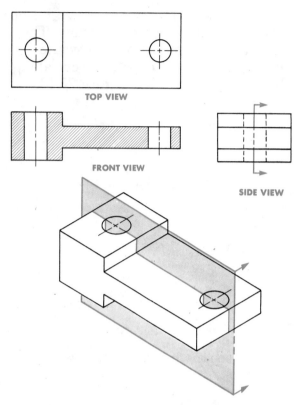

TOP VIEW

FRONT VIEW

SIDE VIEW

Figure 20-9: A full section drawing with the cutting plane line through the side view.

Different types of materials are shown by different line symbols. The line symbols for a few materials are shown in figure 20-6.

A full section is a sectional drawing that shows the whole object cut in half (figure 20-7). A full section can be made from either the front view (figure 20-8) or the side view (figure 20-9). Hidden lines are not shown on a sectional view.

SELF CHECK

1. What is the purpose of a sectioned view?
2. What is the name of the line used to cut an object in half for a sectioned view?
3. What do the arrowheads on the end of the cutting plane mean?
4. What is the angle of the section lines on an orthographic view?

Problems for Unit 20

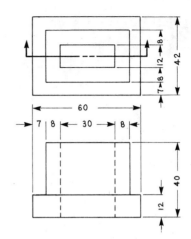

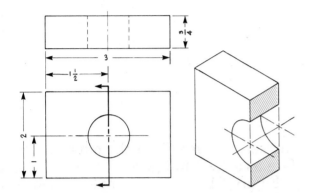

Problem 20-1: Draw the missing full section.

Problem 20-3: Draw the missing full section.

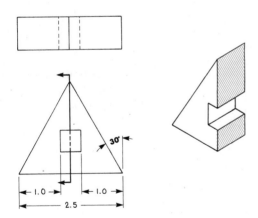

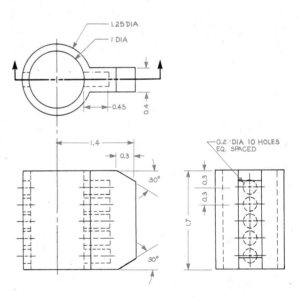

Problem 20-2: Draw the missing full section.

Problem 20-4: Draw the missing full section.

When an object to be drawn is symmetrical, a half section (figure 21-1) may be used. A <u>half section</u> view is a sectional drawing in which the cutting plane cuts through only 1/4 of the object. A <u>center line</u> is used to separate the section view from the outside surface as shown in figure 21-2.

Figure 21-3 is an example of how a half section is developed.

There are other types of sectional views, but the rules for drawing them

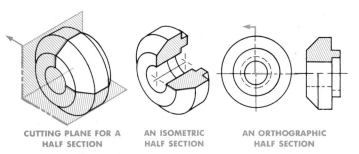

CUTTING PLANE FOR A   AN ISOMETRIC   AN ORTHOGRAPHIC
HALF SECTION   HALF SECTION   HALF SECTION

Figure 21-1: A half section drawing shows the inside of a symmetrical object.

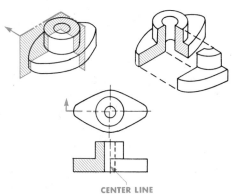

CENTER LINE

Figure 21-2: Note that on an orthographic view a center line is used to separate the half section from the outside view of the drawing.

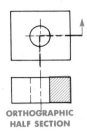

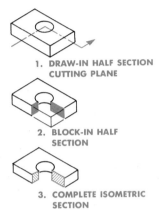

ORTHOGRAPHIC
HALF SECTION

1. **DRAW-IN HALF SECTION CUTTING PLANE**

2. **BLOCK-IN HALF SECTION**

3. **COMPLETE ISOMETRIC SECTION**

Figure 21-3: The steps in drawing a half section.

are complicated. These other types of sectional views will be explained in an advanced drafting course.

SELF CHECK

1. How much of a view is cut by a half section cutting plane?
2. What kind of an object is usually drawn as a half section?
3. Which line separates the sectioned part of the drawing from the outside?
4. What kind of line is used to separate a half section from the outside view on an orthographic drawing?

Problems for Unit 21

Problem 21-1: Draw the missing half section.

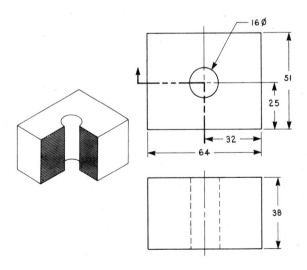

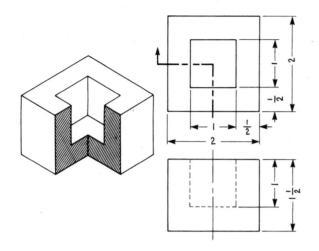

Problem 21-2: Draw the missing half section.

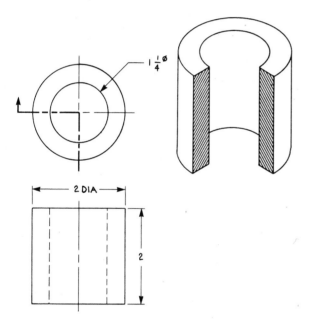

Problem 21-3: Draw the missing half section.

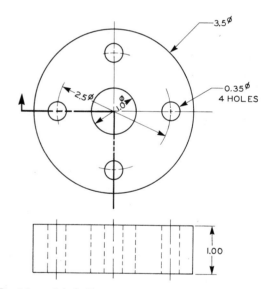

Problem 21-4: Draw the missing half section.

Auxiliary views are additional views added to a regular orthographic drawing (figure 22-1). Auxiliary drawings are usually needed when the object to be drawn has a <u>slanted surface.</u> Regular orthographic drawings, unlike auxiliary drawings, show only the true length of slanted surfaces in the edge view (figure 22-2). Note that in figure 22-3 the auxiliary view is parallel to the slanted surface. The projections are perpendicular (90°) to the slanted surface. In such views, the thickness of the

Figure 22-1: Many additional views of any object are possible.

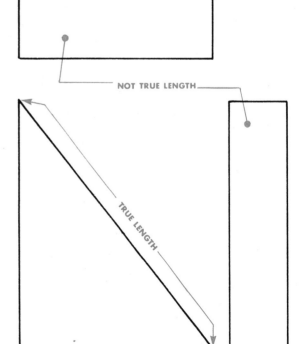

Figure 22-2: Note that the orthographic drawing does not show the slanted surface in true size.

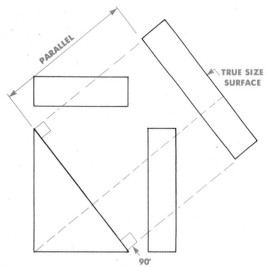

Figure 22-3: The auxiliary view shows the slanted surface in true size.

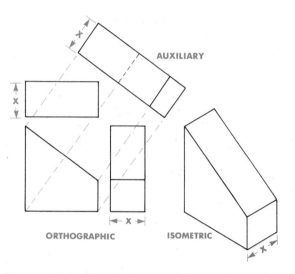

Figure 22-4: The thickness of the auxiliary view is the same as the top and side views.

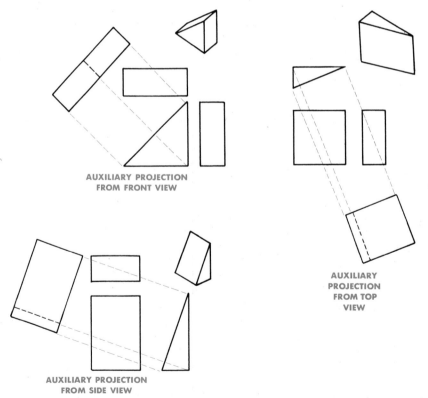

Figure 22-5: An auxiliary view can be projected from any orthographic view.

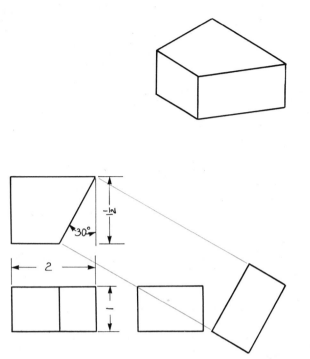

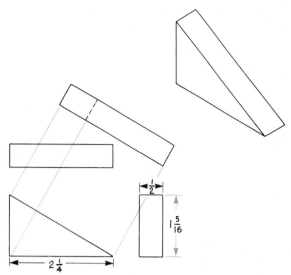

Figure 22-7: An auxiliary projection.

Figure 22-6: This is a partial auxiliary view.

auxiliary is the same thickness as the top and side view of the orthographic (figure 22-4).

Different projections are possible because the auxiliary view can be projected at any angle and from any view of an orthographic (figure 22-5). Thus the auxiliary drawing can show any surface at its true size.

When a full auxiliary is not necessary, only part of the auxiliary view is drawn. This is called a partial auxiliary (figure 22-6). Other auxiliary projections are shown in figures 22-7 through 22-9.

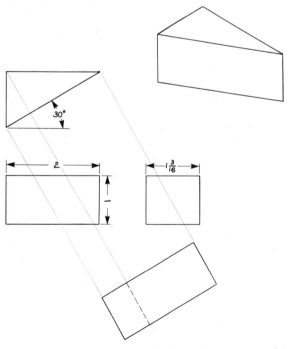

Figure 22-8: An auxiliary projection.

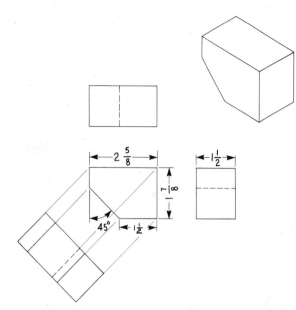

Figure 22-9: An auxiliary projection

1. How many auxiliary views of an orthographic view can be drawn?
2. What is the main purpose of an auxiliary view?
3. What type of surfaces are most auxiliaries drawn from?
4. What is an auxiliary view?

Problems for Unit 22

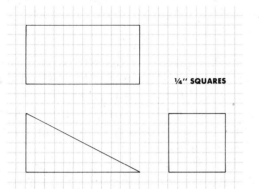

Problem 22-1: Draw the required auxiliary view.

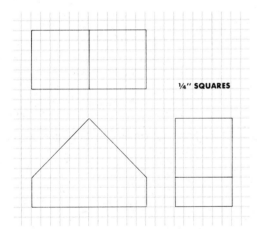

Problem 22-2: Draw the required auxiliary view.

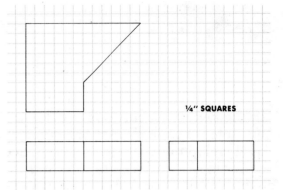

Problem 22-3: Draw the required auxiliary view.

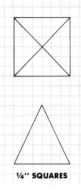

Problem 22-4: Draw the required auxiliary view.

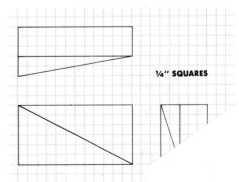

Problem 22-5: Draw the required auxiliary view.

In the next three units you will learn about some of the different types of surface development drawings. Surface development drawings are called "developments" or "pattern developments".

A pattern development is a drawing of the surface of a three-dimensional object drawn on a flat plane (figure 23-1). A pattern is drawn directly on the material to be used. It is then cut and formed to shape (figure 23-2). The four basic development forms are: prism (box), cylinder, cone, and pyramid (figure 23-3).

This unit will explain how to develop the pattern for a rectangular prism. Unit 24 will explain how to develop the pattern of a cylinder, and Unit 25 will

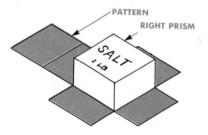

Figure 23-1: A right prism and its pattern.

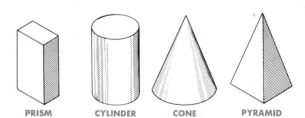

Figure 23-3: The four basic development forms.

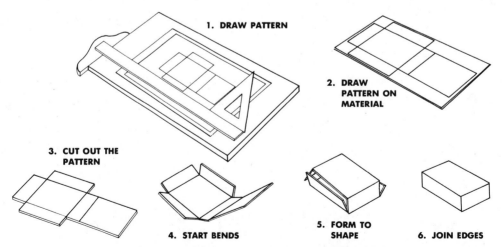

Figure 23-2: The steps in making a box.

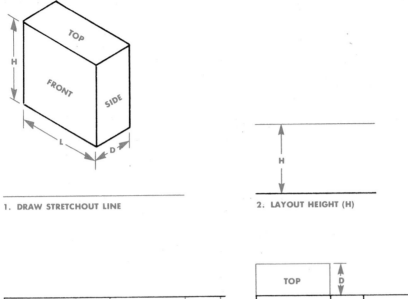

1. DRAW STRETCHOUT LINE

2. LAYOUT HEIGHT (H)

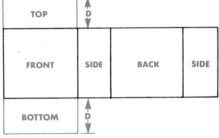

3. STEP OFF FRONT, 2 SIDES AND BACK

4. ADD TOP AND BOTTOM FOR SIX SIDES

5. CUT-OUT, FOLD, AND ASSEMBLE

Figure 23-4: The steps in drawing a prism pattern.

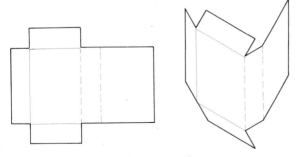

Figure 23-5: Use either a thin solid line or a dotted line as a fold line.

explain how to develop patterns for cones and pyramids.

The steps in drawing a right angle prism (one in which the opposite faces are parallel) are shown in figure 23-4. All lines and surfaces must be drawn full size.

As shown in figure 23-5, the <u>folding lines</u> (those lines on which a pattern is folded) can be either a thin, solid line or a dotted line.

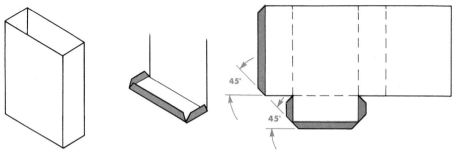

Figure 23-6: A tab seam helps the assembly process.

A <u>tab seam</u> is often used in box developments (figure 23-6). The corners of seam tabs are cut at a 45 degree angle so they will not overlap when they are folded.

1. What type of material is used to make a pattern?
2. What are the four basic pattern forms?
3. What is the name of a commonly used seam?
4. What types of lines can be used to show a fold?

Problems for Unit 23

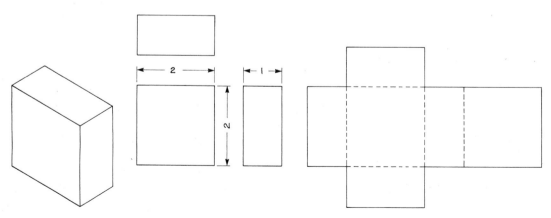

Problem 23-1: Draw the orthographic, isometric, and pattern as assigned by your instructor.

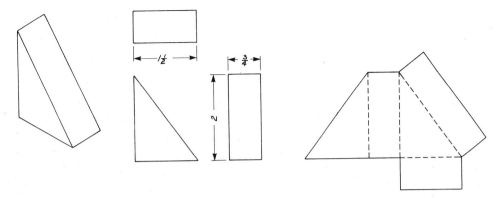

Problem 23-2: Draw the orthographic, isometric, and pattern as assigned by your instructor.

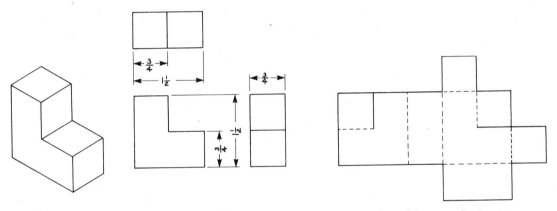

Problem 23-3: Draw the orthographic, isometric, and pattern as assigned by your instructor.

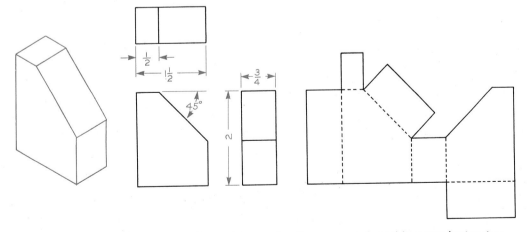

Problem 23-4: Draw the orthographic, isometric, and pattern as assigned by your instructor.

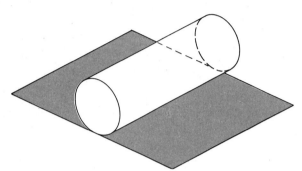

Figure 24-1: A cylinder and its pattern.

In this unit you will learn how to make a pattern for a <u>cylinder</u>. A cylinder (figure 24-1) is one of the four basic development forms. The steps in developing a pattern for a cylinder are shown in figure 24-2. Cylinders may take many different shapes. A <u>truncated cylinder</u>, for example, is shown in figure 24-3.

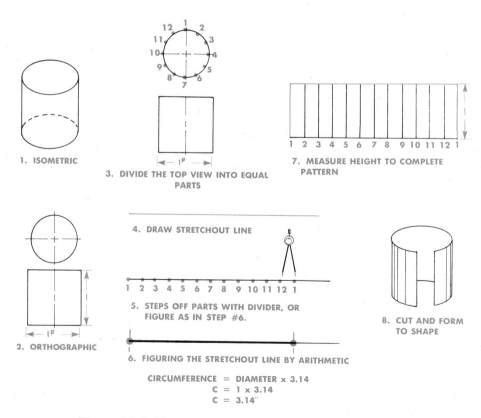

Figure 24-2: The steps in drawing a pattern for a cylinder.

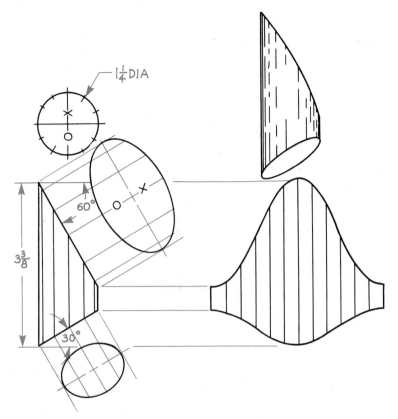

Figure 24-3: The pattern for a truncated cylinder.

SELF CHECK

1. What is the name of the line on which the circumference is laid out?
2. What is the formula to find the circumference of a circle?
3. How many degrees will be in each part of a circle that is divided into 12 parts?
4. What is a cylinder?

Problems for Unit 24

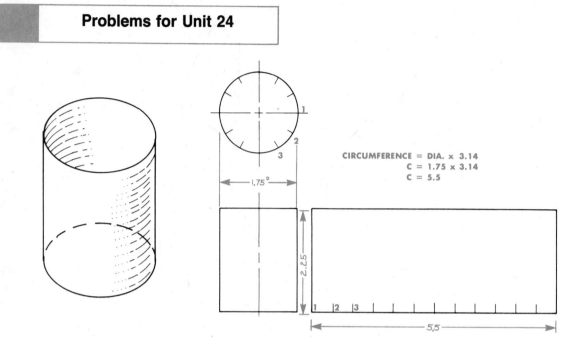

CIRCUMFERENCE = DIA. x 3.14
C = 1.75 x 3.14
C = 5.5

1.75°

2.2.5

5.5

Problem 24-1: Draw the orthographic, isometric, and pattern as assigned by your instructor.

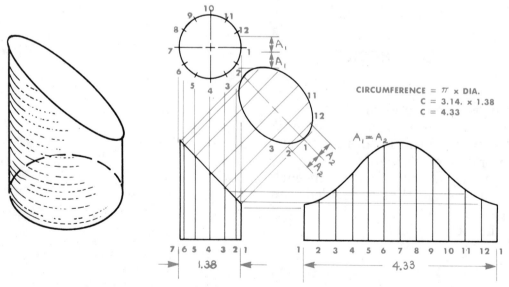

CIRCUMFERENCE = π x DIA.
C = 3.14. x 1.38
C = 4.33

$A_1 = A_2$

1.38

4.33

Problem 24-2: Draw the orthographic, isometric, and pattern as assigned by your instructor.

Freehand sketching is useful for everybody, not just the mechanical drafter. You will, at some time, want to draw a sketch to show something you can't describe in words (figure 25-1). Sketches show ideas to other people and help you remember a good idea. Do you think it would be easier to explain or to show the sketch of figure 25-2?

Figure 25-1: A sketch is better than a thousand words.

Figure 25-2: Would it be easier to explain this space launch in words or to show the space launch in a sketch?

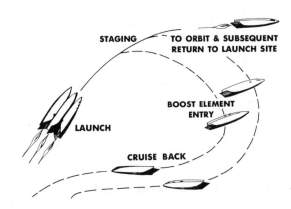

Figure 25-3: Practice sketching horizontal lines.

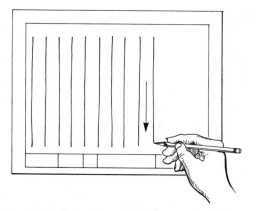

Figure 25-4: Practice sketching vertical lines.

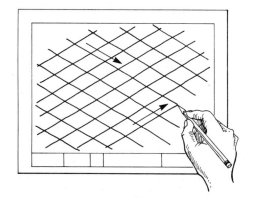

Figure 25-5: Practice sketching angles

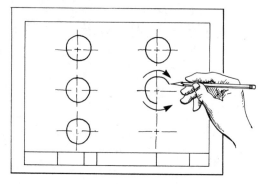

Figure 25-6: Practice sketching circles

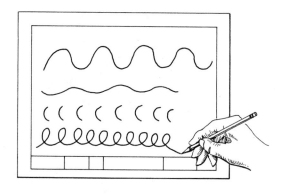

Figure 25-7: Practice sketching curves.

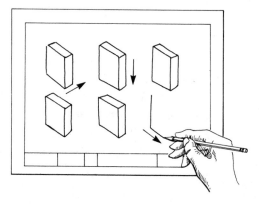

Figure 25-8: Practice sketching isometric blocks.

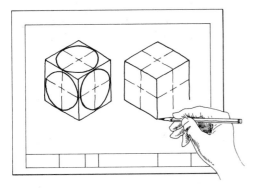

Figure 25-9: Practice sketching isometric circles in isometric squares.

The <u>freehand sketch</u> is the fastest and simplest way to make a drawing. Figures 25-3 through 25-9 are exercises to help you improve your sketching ability.

Types of sketches that are useful include orthographic, isometric, perspective and cabinet (figure 25-10).

Figure 25-11 shows an example of a well drawn and a poorly drawn sketch.

<u>Proportions</u> make a sketch look right. A sketch is not meant to be

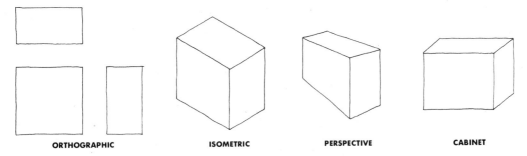

ORTHOGRAPHIC   ISOMETRIC   PERSPECTIVE   CABINET

Figure 25-10: Types of drawings used for sketching mechanical drawings.

Figure 25-11: Good and poor sketches of the same object.

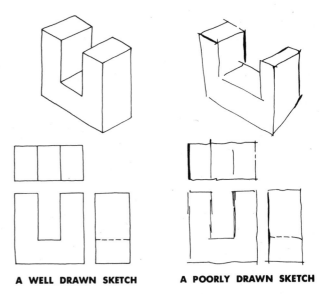

A WELL DRAWN SKETCH   A POORLY DRAWN SKETCH

drawn to exact size. But when the parts of a sketch appear to be in proper proportion to each other, it is called well-proportioned (figure 25-12).

Graph paper helps your speed, accuracy, and ability to make straight lines and estimate proportions whether it has square or isometric lines (figure 25-13).

Sketching offers the following advantages to the draftsperson:
- It helps to develop new ideas
- It shows an idea quickly
- It helps you remember an idea
- It is done more quickly than an instrument drawing
- It is less costly than an instrument drawing

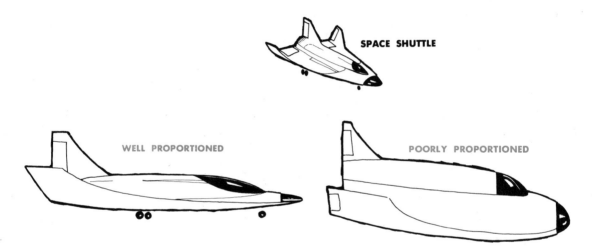

Figure 25-12: Good proportioning for a sketch is important to the overall look.

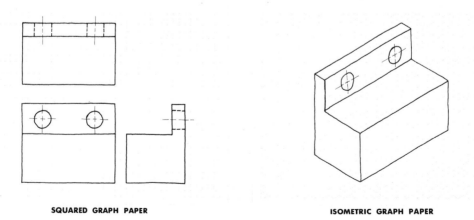

Figure 25-13: Graph paper makes sketching easier.

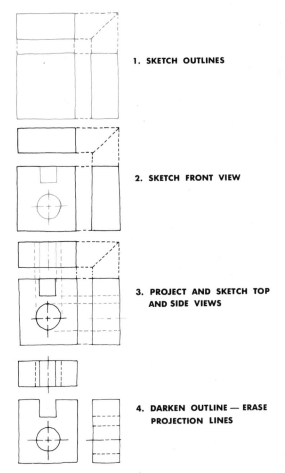

Figure 25-14: Steps for sketching an orthographic drawing.

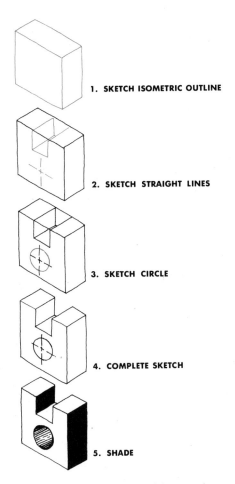

Figure 25-15: Steps for sketching an isometric drawing.

Freehand sketching starts by your choosing the proper (soft) pencil and holding it comfortably; remembering to pull the pencil, not to push it; remembering to turn the paper to draw a difficult line, and practicing constantly.

Steps in sketching an orthographic drawing are shown in figure 25-14. Sketching steps for an isometric drawing are shown in figure 25-15.

SELF CHECK

1. What types of sketches are usually made for industry?
2. What type of paper helps in sketching orthographic drawings?
3. What type of paper helps in sketching isometric drawings?
4. List four reasons why a drafter must know how to sketch.

Problems for Unit 25

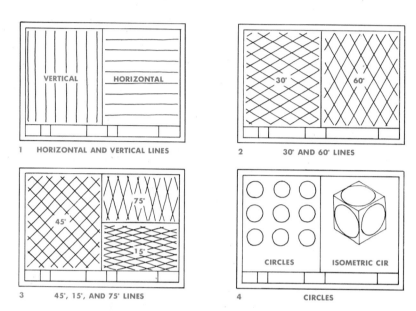

1 **HORIZONTAL AND VERTICAL LINES**

2 **30° AND 60° LINES**

3 **45°, 15°, AND 75° LINES**

4 **CIRCLES**

Problem 25-1: Fill in an "A" size drawing sheet with each of these sketching exercises.

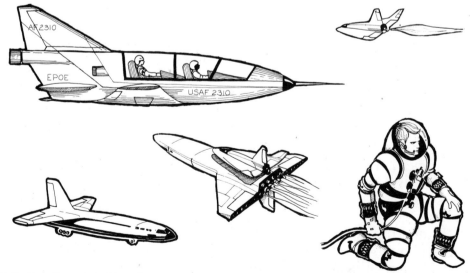

Problem 25-2: Practice sketching these objects as assigned by your instructor.

GLOSSARY

Acute Angle: An angle which is less than 90°.

Allowance: Minimum clearance between mating parts.

Angle: A figure formed by two lines coming together to a point.

Aperture Card: A data processing card containing a cut-out space into which a piece of microfilm is mounted.

Arc: A portion of the circumference of a circle.

Assembly: A unit fitted together from manufactured parts.

Assembly Drawing: A drawing which shows the working relationship of the various parts of a machine or structure as they fit together.

Beam Compass: A compass used to draw arcs and large circles.

Blueprint: A drawing that has been transferred to a sensitized paper by exposure to light and some type of developer, such as ammonia or water.

Bolt Circle: A circular center line locating the centers of holes located about a common center point.

Boss: A cylindrical projection on a forging or a casting.

Casting: Formed by pouring molten metal into a mold of the desired shape and allowing the metal to harden.

Chamfer: A narrow inclined surface along the intersection of two surfaces.

Circle: A closed curve which has all of its points equidistant from its center.

Circumference: The perimeter of a circle.

Circumscribe: Drawing a line around.

Clearance: The distance by which one part clears another.

Component: A single part or unit.

Computer: An electronic calculator which performs a sequence of computations.

Counterbore: The enlargement of the end of a hole to a specified depth and diameter.

Countersink: Forming a depression to fit the conical head of a screw or the thickness of a plate so that the face will be level with the surface.

Cross Hatching: A series of parallel lines which are closely spaced and drawn obliquely to indicate sectional views.

Detail Drawing: A drawing of a single part which provides all the information which is needed in the production of that part.

Development: The pattern of the surface of an object drawn on a flat plane or surface.

Diameter: The length of a straight line which runs through the center of a circle.

Drill: Making a cylindrical hole using a revolving tool with cutting edges.

Ellipse: A Closed curve which is in the form of a symmetrical oval.

Engineering Drawing: Also known as a technical drawing or drafting. It is the graphic language of the engineer.

Exploded View: Separate parts of a single assembly projected away from each other, or separated showing relationships among the parts of a drawing.

FAO: An abbreviation used on detail drawings to indicate that the piece is to be *finished all over*.

Fillet: An interior rounded intersection between two surfaces.

Flange: A projecting rim which adds strength, provides for an attachment to another part, or acts as a guide.

Hexagon: A six sided figure with each side forming a 60° angle.

Knurl: Impressing a pattern of dents in a turned surface with a knurling tool to produce a better hand grip.

Line Conventions: Symbols which represent or describe some part of an object. It is expressed by a combination of line weight and appearance.

Microfilm: A high resolution photographic film, usually 35 mm size for engineering drawings and documents.

National Coarse (NC): The coarse thread series of the American Standard screw threads.

National Fine (NF): The fine thread series of the American Standard screw threads.

Nominal Size: A general classification term which designates the size of a commercial product.

Object Line: The heavy, full line on a drawing which describes the object.

Obtuse Angle: An angle greater than 90 degrees.

Octagon: An eight-sided geometric figure. Each side forms a 45 degree angle.

Orthographic Projection: A multiview drawing which shows every feature of an object in its true size and shape.

Over-all Dimensions: Dimensions giving the entire length or width of an object contrasted with dimensions showing small details of an object, such as the location of a hole.

Pentagon: A five sided geometric figure, each side forming a 72 degree angle.

Perimeter: The boundry of a geometric figure.

Perpendicular: A line which is at right angles to a given line.

Perspective Drawing: A method of pictorial drawing representing an object on a single plane as it appears to the eye.

Radius: The length of a straight line which runs from the center to the perimeter of the circle.

Rib: See Web.

Right Angle: A 90 degree angle. The angle formed by a line that is perpendicular to another line.

Rotate: Turning or revolving around a point.

Round: The rounded over corner of two surfaces.

Section: A cross sectional view at a specific point of a part or assembly.

Section Lining: Thin lines usually drawn at a 45° angle to the horizontal. They represent the internal parts of an object as if part of the object had been cut away.

Shaft: A cylindrical piece of steel which is used to carry pulleys or to transmit power by rotation.

Shoulder: A plane surface on a shaft, normal to the axis and formed by a difference in diameter.

Sketch: Drawing without the aid of drafting instruments.

Spot Face: Finishing a circular spot slightly below a rough surface on a casting to provide a smooth, flat seat for a bolthead or other fastening.

Stretch-out: A full size drawing or pattern of a sheet metal object.

Symbol: A figure or character which is used instead of a word or a group of words.

Tangent: A line drawn to the surface of an arc or a circle so that it contracts the arc or circle at one point only.

Taper: A piece that increases or decreases in size at a uniform rate and assumes a wedge or conical shape.

Technical Illustration: A pictorial drawing made to simplify and interpret technical information.

Tolerance: The total amount of variation allowed from the design size of a part.

Truncate: Cutting off a geometric solid at an angle to its base.

Washer: A ring of metal used in forming a seat for a bolt or nut.

Web: A thin flat part which joins larger parts. It is also known as a rib.

ABBREVIATIONS

Abbreviations are short forms of words. Their use in drafting saves time and space on the drawing. Generally, abbreviations used on drawings require no periods, except those abbreviations that spell words.

Aluminum	AL
And	&
Assembly	ASSY
Auxiliary	AUX
Bearing	BRG
Bill of Material	B/M
Bushing	BUSH.
Casting	CAST
Cast Iron	CI
Center Line	
Centimeter	cm
Chamfer	CHAM
Cold Rolled Steel	CRS
Counterbore	CBORE
Countersink	CSINK
Deep	DP
Degree Celsius	°C
Detail	DET
Diameter	D
Dowel	DWL
Drawing	DWG
Drawn	DWN
Drill	DR
Each	EA
Finish	FIN
Finish All Over	FAO
Fixture	FIXT
Gage	GA

Grind	G,GRD
Harden	HDN
Head	HD
Hexagon	HEX
Holes	HLS
Hot Rolled Steel	HRS
Inch	″ IN.
Inside Diameter	ID
Kilometer	km
Liter	l
Machine Steel	MS
Manufacture	MFR
Material	MATL
Maximum	MAX
Meter	m
Millimeter	mm
Nominal	NOM
Number	NO.
Outside Diameter	OD
Pattern	PATT
Press Fit	PF
Plus or Minus	±
Quantity	QTY
Radius	R
Ream	RM
Reference	REF
Reproduction	REPROD
Required	REQD
Screw	SCR
Section	SECT.
Slip Fit	SF
Standard	STD
Steel	STL
Symbol	SYM
Thread	THD
Through	THRU
Tolerance	TOL
Typical	TYP
Volt	V

INDEX